SpringerBriefs in Ethics

For further volumes:
http://www.springer.com/series/10184

Lisa Newton

The American Experience
in Bioethics

 Springer

Lisa Newton
Shelburne, VT
USA

ISSN 2211-8101   ISSN 2211-811X   (electronic)
ISBN 978-3-319-00362-7   ISBN 978-3-319-00363-4   (eBook)
DOI 10.1007/978-3-319-00363-4
Springer Cham Heidelberg New York Dordrecht London

Library of Congress Control Number: 2013936002

Printed on acid-free paper

Springer is part of Springer Science+Business Media (www.springer.com)

Contents

Introduction

This piece is an introduction to the ethical problems surrounding the provision of health care: how the problems present themselves, something of the history of our attempts to deal with them, whatever consensus may have been achieved—or not, depending on the problem. Given the breadth of the topic, no pretense has been made to go into any part of it in depth; given the mission of this series for brevity, all footnotes/endnotes have been omitted. Where facts and figures would be relevant (for instance, in the number of U.S. citizens not covered by healthcare insurance, or the cost of that insurance), those facts and figures have generally been omitted, since they go out-of-date very quickly (and are readily available electronically). This account is meant to be no more than a Brief, or briefing—an introduction to the controversies for those who need to know something about them in small compass.

The first two chapters track the development of Bioethics in America in an order dictated by the importance of the issues in traditional discussions of the subject. First, birth and death, abortion and euthanasia, and the issues that spread out from them: Do couples anxious for a pregnancy have a right to technological assistance? If we want a certain type or appearance of the child, should we meddle with the genes to try to get the perfect baby? Can we ensure genetic continuity by cloning a parent? Once a pregnancy is established, do we have a right to terminate it? What if the child is born deformed or missing organs necessary for life? Are there lives that are simply not worth living? What do we mean by "death with dignity"? What are we permitted to do to protect the dignity of the dying? Second, when biomedical practice steps into the realm of the experiment—using human beings as its guinea pigs—how can we make sure that the subjects' rights as human beings are preserved? Is there "experimentation with dignity"? Are cutting edge practices—organ transplantation to begin with, now transplantation of entire limbs—likely to do more harm than good? Who can be the judge of these practices? Recall, it was outrage over research with human subjects that led to the formation of the National Commission for the Protection of Human Subjects of Biomedical and Behavioral research, and the best public discussion of the new challenges to Bioethics that our country has enjoyed.

After this necessarily brief tour through the emotionally searing cases that shaped what consensus there is on Bioethics in America, we take on the great conundrum of our time: the provision of excellent health care for all Americans at an affordable cost. This proposal is new; it has been aired in a few specialized journal entries and a few academic conferences, but I have not seen its like taken seriously in the current political debate. Its adoption would require the upending of many of our assumptions about healthcare practice—but not all of them. Parts of the proposed system are in place now, and others would not be difficult to restore. But it will entail a vast change in the patterns and amounts of compensation for the provision of health care, and that change will create enough political problems to keep us occupied for some decades. We had best get started immediately.

Chapter 1
Birth, Life, Death

From the beginning of the passionate discussions of the ethical implications of the human condition, especially the condition of the human body and our ultimately futile efforts to avoid disease and stave off death, we have continually returned to the basic questions of life and death: if life is of immeasurable value, can there possibly be conditions where it would be right to cut it short, to end it? No course in medical ethics[1] can avoid the questions of abortion and euthanasia, the termination of life before it really begins or hastening its end as it approaches. Can these be morally justified? On examination, the questions expand to include all manipulation of the processes of human life, from the most basic cells from which we emerge, through their growth to birth, the technologies we have developed to assist reproduction, through the perils of human life to the approach of death. What right have we to shape these natural processes for our comfort and convenience? These most basic questions shape the beginnings of the discipline.

1.1 Issues Surrounding the Basic Material of Life

Ought we to isolate and manipulate the stuff of life itself—including research with embryonic stem cells, experiments with cloning human beings, creating chimeras to discover new possibilities for tissue transplantation, trying genetic modification of human embryos to eliminate undesirable genetic traits (or enhance desirable ones)—even to achieve worthy medical, personal, or social goals? In all these cases, the "greatest happiness" of the people at large will be served by continuing the research. But is it compatible with the dignity of the human?

---

[1] Since "medical ethics" seems to apply to physicians only, more recent texts name the field "health care ethics," which applies to all the health care fields, or "bioethics," which includes biomedical and behavioral research as well as treatment of patients.

L. Newton, *The American Experience in Bioethics*, SpringerBriefs in Ethics, DOI: 10.1007/978-3-319-00363-4_1, © The Author(s) 2013

1.1.1 Stem Cell Research: Some Definitions

Stem cell: an undifferentiated cell, from which an organism develops. Ordinarily found at the beginning of life, although some persist into the adult body to become replacement cells.

Totipotent stem cell: a cell that has the potential to develop into a complete organism.

Pluripotent stem cell: a stem cell that has the potential to differentiate into any of the three germ layers of ectoderm, mesoderm, or ectoderm, but not gametes, so lacks the potential to develop into a complete organism.

Human embryonic stem cell (hES cell): pluripotent stem cell derived from the morula stage of the embryo, between 7 and 14 days after conception.

Induced pluripotent human stem cell (iPS cell): a pluripotent stem cell that has been artificially created by reprogramming non-pluripotent human cells through techniques that do not involve oocytes or embryos, e.g. through inserting certain genes into a somatic cell.

Why would we want to do research with stem cells?

The study of any aspect of life is a worthy project in itself. But in the most important projected application, we may be able to develop lines of cells that can be placed in an impaired patient to repair serious damage that we cannot treat in any other way. For instance, these cells may be able to turn themselves into nerve cells in the severed spine of a paralytic, to connect the torn ends of the spinal cord and restore function lost in the injury. (Nerve cells are the only cells in the body that normally do not regenerate.)

> **Superman**: Christopher Reeve (1952–2004), a professional actor most famous for his role as Superman (or Kal-El or Clark Kent) in a series of movies featuring the 1940s comic-book hero, was rendered quadriplegic by a fall from a horse in 1995. Refusing to accept his condition as permanent and incurable, he quickly became a new kind of hero for the nation in his courage in his infirmity, his work for cures for spinal cord injuries, and for the Christopher Reeve Foundation and the Reeve-Irvine Research Center that he and his wife founded for that research. Reeve was very public in support of stem cell research, which he was convinced would come up with a way to heal his spine, and strongly supported this research until the end of his life. Reeve is not alone; such injuries are rare in peacetime, but not in war, and our veterans' hospitals have whole wards of Americans for whom we have no other help. Any research that holds out hope for wounded soldiers— young men with their lives before them—has ample justification, in welfare and in justice.

Cells placed in a brain may be able to repair the brain damage caused by seizures, also Parkinson's Disease and possibly even Alzheimer's Disease.

> **Alzheimer's Disease, strokes, brain injuries, seizures, and other insults to the brain:** There are several degenerative diseases that attack the aging brain, of which Alzheimer's Disease is only the poster child; traumatic brain injury is one of the more common horrors of the current wars, and hemorrhages in the brain (strokes) accompany the increasing affluence (and high blood pressure) of the people of the modern world. All of them destroy brain tissue, which, as above, cannot regenerate. And they are all increasingly troublesome: soldiers who would have died in all previous wars are being saved for a life of brain impairment; diseases of age are increasing as the population ages; strokes that

used to kill now only injure. Could pluripotent stem cells, placed in the brain of one suffering from one of these conditions, take on the character of the brain in which they are
placed, grow to replace the neurons lost to disease or trauma, and restore the damaged
brain?

Considering the immense suffering such diseases cause, to those who suffer from
them and possibly even more to the caregivers, any research that can hold out any
promise of relief has a strong utilitarian warrant. Then why would such research
be controversial? Because not all parties to the debate are utilitarians. Consider: If
we are dealing with embryonic stem cells, the research requires growing a human
embryo for about a week, then destroying it to harvest the stem cells. But many
people believe that the humanity of the human person begins with its biological
beginning, at conception; that the fertilized egg (and *a fortiori* every stage of the
embryo thereafter) is a full human being with full human rights. In that case, the
act of destroying the embryo to harvest the cells is the killing of a human being,
or, in the eyes of those who define the embryo this way, murder.

Most would disagree with this evaluation, but the duty of respect for human life
at any stage of its existence may demand at least that the destruction of embryos
not be casual and wanton. If, given the possibilities for relief of the terrible suffering of paralysis or dementia, it would be irresponsible to abandon stem cell
research, at least, as the research continues, it should be regulated, watched, just
to make sure that unforeseen abuses are not taking place. It should be noted that
the embryo that is so destroyed is usually a frozen embryo left over from In Vitro
Fertilization attempts, after the donor couple has borne a child and has donated the
embryo for research. Should it not be used for research, it will be destroyed.

The same question does not arise, of course, if we are dealing only with skin
cells treated to the point of pluripotency, but a question has already arisen about
the possibility that further manipulation might return such cells to totipotency—
the condition of the initial fertilized egg. In that case, would the manipulated skin
cell become a human being? The thought is disquieting.

1.2   Cloning: Ought We to Attempt to Clone Humans?

What is a "clone"? Strictly speaking, a "clone" is a group, any number of genetically identical organisms, like identical twins or triplets. We tend to use the word
to mean a single organism, demonstrably genetically identical to another. Clones
may occur naturally, as in normal identical twins. Or they may be constructed by
mononuclear reproduction, extracting the DNA from a somatic cell of the parent
organism, enucleating a viable egg cell from the recipient, and placing the DNA
into the egg to create a complete being, which inserted into the recipient's womb,
then goes on to develop normally.

The possibility of "cloning," mononuclear reproduction, was first raised in
the early part of the twentieth century by German embryologist Hans Spemann,
who had received the 1935 Nobel Prize for Medicine for his work in embryonic

development. By the 1960s, molecular biologists were experimenting with frog eggs, and through the 1970s there were claims of successful cloning of frogs and mice, widely suspect. It's worth noting that even the earliest work on cloning was accompanied by claims of success and heated accusations of fraud; this type of work has never been non-controversial.

Meanwhile, the laboratory techniques were improving; in 1978 the first "test-tube baby" (Louise Brown, see below) was born. A lamb cloned from embryonic cells emerged from Cambridge, England, in 1984, calves from Wisconsin and sheep from the Roslin Institute in Scotland followed in the 1990s, and finally, in 1997 from the same Roslin Institute, came Dolly.

> **Dolly**: Ian Wilmut and his colleagues, building on the earlier work of Steed Willadson in Texas, claimed the first lamb cloned from a fully differentiated cell (taken from her mother's udder). She was named "Dolly," for Dolly Parton, since the cell had been taken from a breast. The breeders delivered the lamb on July 3, 1996, but waited seven months to announce it in February 1997, to make sure that the patents on the processes were granted. There are several ways to produce clones; Dolly was born of Somatic Cell Nuclear Transfer (SCNT), which takes the nucleus of an adult cell and implants it in an egg cell where the nucleus has been removed, or "fuses" the somatic cell with an enucle-ated egg cell by a small electric current. (The two procedures result in slightly different complements of mitochondrial DNA; Dolly was born of the second.) The birth caused an enormous stir, although Ian Wilmut was the first to point out the vast inefficiency of the process: he had started with 277 eggs fused with sperm, recovered 247 of them, trans-ferred the healthiest 29 at the blastocyst stage to sheep wombs, resulting in 13 pregnan-cies, three of which came to term and one of which survived. Dolly was anxiously cared for, kept away from sources of disease, but developed arthritis early in life and died at a very young age from a common degenerative condition.

That was Dolly, and there are easier ways to get lambs or the shepherds would have been out of business centuries ago. Moreover, in a major disappointment for those who had been dreaming of creating humans identical to themselves, Dolly did not even look like her mother: her mother's face was black, Dolly's was white. (The same result occurred in the first attempt to clone a cat, carefully watched by all Tabby-lovers who could not bear the thought of separation from their pets—the "clone" was calico, all right, but in a totally different pattern of spots.) The sci-entists pointed out that clones constructed this way are *not* identical. Contrary to popular belief, the nuclear DNA, contained in the transplanted nucleus, is not the complete genetic blueprint of the individual. Nuclear DNA provides only some of the genes needed for a complete organism. The rest of the genetic material is sup-plied by the egg itself, the mitochondrial DNA outside the nucleus. So Dolly was not "identical" with her mother, along several dimensions, and most of the clones since (mice, a cat or two) are distinctly different from their sources. Meanwhile, an indefinite amount of congenital traits (characteristics present when the organ-ism is born) are determined by the prenatal environment in the womb.

When news of Dolly broke in the news media, the U.S. Congress leaped into the breach with a law forbidding (the constructing of) human clones. The alarm seems premature. After all, why would we *want* to clone humans? That's hard to say; many considerations just to this point could lead us to decide that there is no reason at all to attempt to clone a human. First, it's an awful lot of trouble (and expense) for very small returns. Dolly took 277 tries, as above, and at the end, we didn't have "the same" sheep at all. Second, the clone may have inherited more trouble than identity from its origins. From the early death of Dolly, we have reason to believe that an artificially constructed clone may be not an offspring, an identical child, but a late-born twin, physiologically the same age as the source, and destined for "early" illness and death. (The age of any organism is governed by the length of certain structures in the nucleus, called "telomeres." With each division of the cell, the telomeres get shorter, and when they are gone, the cell can divide no more and aging is in process. Was Dolly born with the aging telomeres of her mother?)

Why is cloning so controversial? What on earth was the U.S. Congress afraid of, when they rushed to pass anti-cloning legislation? The question has some interesting answers. While cloning turns out to be difficult, uncertain, and disappointing in its results in the real world, it has formed the stuff of fantasy and science fiction for a long time.

> *The Boys from Brazil*: One of the first works to invite the public to contemplate the cloning of a human being was a fictional thriller, *The Boys from Brazil,* written by Ira Levin in 1976. Levin's previous works had included *The Stepford Wives,* about a community that replaced its uppity women with agreeable robots, and *Rosemary's Baby,* about a woman impregnated by the Devil; the author was at home with horror. In *Boys,* the morally problematic activity of producing a clone at all is combined with the morally abhorrent motivation of the ultimate success of the Aryan Project and the memory of the Holocaust, as the plot involves Dr. Josef Mengele (alive and well and living in Paraguay) in a scheme to produce 94 replicas of Adolf Hitler. The story has Mengele making it into exile with a liter of Hitler's blood, and hiring willing native women to contribute eggs and become impregnated with the cell bearing Hitler's DNA. When born, a cooperative ex-Nazi file clerk in an adoption agency in New York finds adoption applicants that match the profile of Hitler's parents, and the baby boys are raised by these adoptive parents (sworn to secrecy about the fact of adoption) until the age of 14. Then, unfortunately for them, the adoptive fathers have to die, since that's when Hitler's father died, and that means 94 murders. The story begins as the murder list is drawn up in a Brazilian bar by Nazis in South American exile, and ends (in an enormously satisfying scene) with Mengele being eaten by Dobermans.

The sheer evil of demonic invention that permeates the tale is punctuated with factual errors, no fault of Levin's: the story assumes that the procedure guarantees pregnancy and normal birth almost every time, making the terrifying scenario seem to be just around the corner. Further, it assumes that the clone is in fact identical in all respects to the source of the nuclear DNA, just as a metastasized cancer cell is the *same tumor* as the original—that each of the boys *is literally Hitler.* (The protagonist, a Jew, a Nazi-hunter on the model of Simon Wiesenthal, has great difficulty bringing himself to shake the hand of one of the boys, seeing him *as Hitler*; he deflects only at the last minute a further plot to kill all 94 of the children, a plot born in fear of a high probability that any one of them would duplicate Hitler's career.) Is it possible that such works as *Boys,* or *Boys* itself (it was a best-seller), is the source of the moral panic about "cloning"? There is little else to account for it; our experience of identical twins, who share mitochondrial as well as nuclear DNA, the same prenatal environment and usually the same upbringing, dressed in identical sailor suits for the first twelve years of life, is that they rarely turn out to engage in identical careers or other behavior. Clearly the *Boys* project of duplicating all the conditions of Hitler's boyhood—including World War I?—is silly fantasy. But some horrors strike deep in the human soul, and are hard to dislodge.

South Korea took an early lead in cloning research (later discredited by over-ambitious false claims), succeeding in 1999 in deriving a four-celled embryo from a somatic cell of an infertile woman; given the controversies, they did not implant the embryo. Meanwhile teams of researchers in Massachusetts, Japan, and Scotland cloned goats and pigs with modified DNA—the goats modified to produce special proteins in their milk, and the pigs modified for greater compat-ibility with human tissue, to be used for tissue replacement. In 2000, after many tries, a team at the Health Sciences University in Portland, Oregon, led by Gerald Schatten, managed to create a clone of rhesus monkeys the old-fashioned way—not by evacuating the nucleus of an embryo and inserting a somatic nucleus, but by growing an eight-celled embryo, splitting it manually into four parts, and implanting each into the womb of a receptive female. "Embryo-splitting," or divi-sion of the original cell mass, is the way twinning (or in this case, quadruplet-ing) would naturally occur in the womb of a pregnant female, and the process was therefore less complicated and prone to failure than the mononuclear reproduction techniques that gave us Dolly.

1.2.1 Experiments with Stem Cell Therapy

While cloning may not have much of a future, the use of stem cells to replace impaired cell systems in seriously impaired patients seems to hold out much more hope. What can we expect from stem cell therapy? As it happens, we can watch current experiments, for there are claims that it is being done now.

> **Stem Cell Therapy (***Quackwatch***)**: A website named "quackwatch," that monitors medi-cal scams and illicit claims, has a long list of "medical centers" around the world which promise miraculous cures for everything from A to Z—ALS, birth defects, cancer, dia-betes, epilepsy and so on. The tragic incurable wasting and degenerative diseases, like Parkinson's and muscular dystrophy, are most frequently mentioned—precisely those dis-eases that kill slowly, horribly, and prematurely, and for which modern medicine holds out no hope of cure. Several of these have had headquarters briefly in the US, but tended to have their materials and equipment seized as the FDA determined that they were scams. These snake-oil establishments operate most freely in China, the Ukraine, and the Dominican Republic and other locations in the Caribbean. (It is pointless to list the "cent-ers" and their locations—they move constantly, since continual relocation is their best defense against national monitors attempting to reduce the number of medical scams that prey on the false hopes of their citizens.)

Some interesting results have been achieved inserting certain kinds of stem cells in animal models. But there is no evidence, anywhere, that any of these experimental therapies are doing any good at all for humans. Most of them do no harm, but signature false claims (for instance, that their stem cells do not trigger an immune reaction in the recipient, or that cells injected in the abdomen inevitably find their way to the brain and become nerve cells, or that the amount of cells inserted makes no difference) reveal that the treatments lack any scientific basis.

1.3 Assisted Reproduction

Ought we to manipulate the process of conception and birth in order to help infertile couples have a baby? Are the various techniques of assisting reproduction (In Vitro Fertilization, AIH, AID, egg donation, "surrogate motherhood") compatible with the integrity of the natural process of conception and birth? No doubt there is some risk, and inevitable financial burdens, to the parties who participate; how seriously should that be weighed, as long as informed consent is obtained?

1.3.1 Technology of Assisted Reproduction

If a couple wants a baby, and the usual way doesn't seem to be working, what means are available to raise the probability of a healthy pregnancy? First, the couple needs to find out if there is some medical problem—something in them blocked or not functioning—that is causing the problem, and whether or not it can be cured. The path to producing a healthy ovum (egg) in the woman's body, for instance, is very complex, starting with hormones released by the hypothalamus in the brain. The path to producing healthy sperm also starts in the hypothalamus and involves multiple steps, failure of any one of which will prevent the ejaculation of sperm into the vagina. The egg has to be released from the ovary, and travel down into the fallopian tube; the sperm has to penetrate the cervical mucus, move quickly through the uterus and meet the egg in the fallopian tube. If there are blockages at any point in this journey, fertilization will not happen "naturally," and if there is no evident remedy, alternative methods must be explored. Let's take these one at a time.

Remedying medical conditions: If the fallopian tubes, for instance, are blocked by disease, they can be cleaned out. The process is not easy, and does not always work, but if it does, a natural pregnancy becomes possible. There are other physical conditions that may count as illnesses, or handicaps, that stand in need of remedy. Those who (like the theologians of the Roman Catholic Church) hold that all "reproductive technology" is wrong, have no objection to any medical interventions that relieve illness, disease, or handicap.

After that, anything done not to relieve infirmity, but to promote pregnancy, counts as reproductive technology.

AIH, Artificial Insemination by Husband: If it is determined that the major obstacle to pregnancy is some fault in the process of copulation itself, the husband's sperm can be collected, concentrated, and introduced back into the wife's vagina when subtle temperature changes indicate that she is ovulating. Sometimes this works.

AID, Artificial Insemination by Donor: If the husband turns out to be infertile, donor sperm is available in sperm banks maintained by medical schools. Donated sperm, screened for disease and physical resemblance to the couple, may be introduced into the vagina as above.

In Vitro Fertilization: When conception cannot take place within the body of the woman, the wife's eggs can be retrieved by laparoscopy to be mixed with the husband's sperm and hormones in a Petri dish, where conception can take place. This is how the first "test-tube baby," Louise Brown, was conceived. The resulting zygote can be inserted directly into the uterus (although current practice attempts to place it up the fallopian tube, if the tube is undamaged, where normal conception would occur). Estimates vary (and should be taken with a grain of salt), but many thousands of babies have been born from IVF—over a million from some kind of artificially assisted reproduction technology (ART). Still, of those who attempt to conceive through IVF, three quarters will not, in the end, have a baby.

Egg donation: The greatest obstacle to pregnancy is the age of the egg, not the age of the gestator (the woman who carries the fetus and gives birth to the baby). A small industry has grown up retrieving the eggs of younger women for donation to women over 40 who still want a child. Given the demand, there is no end in sight for this procedure, or market.

Surrogate "motherhood": If the woman is simply unable to conceive or bear a child, one alternative for the couple is surrogate motherhood, in which the sperm of the husband is introduced into a fertile woman contracted for the purpose, who will carry the fetus to term, deliver it and transfer it to the contracting couple. This practice has been particularly controversial, since there is a strong hormonal motive for the surrogate to bond to the child, and find it very difficult to give the child up to the couple that hired her services. (See *Mary Beth Whitehead,* below)

Surrogate gestation: If for any reason the gametes of both husband and wife are not desirable for reproduction, and the wife is unable to bear a child (if disease has required a hysterectomy, for instance) the couple may avail themselves of an egg bank for a young and healthy egg, a sperm bank for appropriate healthy sperm, and the services of a healthy surrogate gestator, to receive the implanted embryo and bring the child to term. In this case, the child may have five "parents": egg donor, sperm donor, gestator, and the couple that originally contracted for the child.

The position of the Roman Catholic Church: Essentially, that God created procreation, and it doesn't seem to need any help, so we should keep our cotton-picking fingers off the whole process.

Louise Brown: The first baby born as a result of **In Vitro Fertilization (IVF)** was Louise Brown, who came into the world on July 25, 1978. Her birth was the result of many, many years of research. Physiologist Robert Edwards had spent decades learning how to bring about "superovulation," the production of multiple ova, first in mice, then in humans, so that the many mature eggs needed to increase the chance of fertilization might be available. Patrick Steptoe, a gynecological surgeon, had been working, not with Edwards, for the same amount of time on techniques to retrieve fertilized eggs through laparoscopy (a new technique using fiberoptics), to bring them outside the body so they could be fertilized in vitro. Realizing they could help each other, they joined forces, but at the time, the late 1960s, such collaboration was very difficult, which may be one reason why these techniques were not developed earlier. (At the time they were working, the communication between their laboratories was a grueling eight-hour commute, and Edwards' lab lacked hot running water.) They recruited 100 infertile women to be the subjects for the new IVF technique, mixing the husband's or donated sperm with eggs retrieved from the wife. Lesley Brown, whose fallopian tubes had become blocked from ectopic pregnancies, was one of these. When all 100 attempts failed, they kept going, still tinkering with the hormone balance. On the 102nd try, Louise was conceived, and was born healthy. (Now an adult, she is still healthy.)

Surrogate Motherhood: In "surrogate motherhood," where one woman agrees to conceive and bear a child for another using the other's husband's sperm, just whose child is the baby when it is born? Consider Noel Keane's innovative practice:

Noel Keane, Mary Beth Whitehead and the Sterns: In 1976, Noel Keane, a lawyer in Dearborn, Michigan, negotiated and drafted the first formal contract between a surrogate mother and a married couple in the United States. The practice caught on: by his death in 1997, Keane counted 600 children around the world whose births he had arranged. Many were named for him by grateful parents. In the seven years after he had plunged into a largely unknown area of law by drawing up his first surrogate-parenthood contract, Mr. Keane rapidly became the best-known lawyer in the field. In 1983, against a background in difficulty in adoption and a scarcity of white babies available for adoption, Mr. Keane said he had been contacted by more than 2,000 couples. He opened Infertility Centers in regions like California, Indiana, Michigan, New York and Nevada to bring together childless couples and prospective mothers. In the years afterward, in the midst of courtroom battles and philosophical, moral, religious and legislative arguments, he remained a strong proponent of surrogate motherhood.

The practice was not new; the book of Genesis describes Abraham's decision to produce a son by his wife's handmaid, when his wife was apparently infertile. From the beginning, the practice was controversial; proponents depicted surrogate parenthood as a humane and ethical way to allow infertile people to reproduce, while opponents called the technique an abuse of scientific technique that equated life with a product. They called it "baby selling."

In a typical "surrogate motherhood" agreement, a woman is artificially inseminated with sperm from a man whose wife cannot conceive. At term, the surrogate mother surrenders her parental rights, receives a fee and turns over the baby to its biological father, whose wife may then become its adoptive mother. In a typical arrangement 20 years ago, the child-bearing woman received $10,000 (rates varied). The fees included $7,500 to the center, as well as charges for legal, medical and travel services and paternity tests. All told, couples in the 1980s and 1990s were advised to expect to pay $22,000 to $25,000.

Although other cases raised issues about surrogate parents, the furor reached a crucial point in 1986 in a trial in New Jersey that grew out of a surrogate motherhood arranged by Mr. Keane. Mary Beth Whitehead signed a $10,000 contract to carry the baby of William Stern. But after the child was born, Mrs. Whitehead changed her mind and sued for custody of the child, known as Baby M. Mr. Stern and his wife, Elizabeth, won custody of the girl, but the State Supreme Court in New Jersey ruled that commercial surrogate-motherhood contracts were illegal and awarded Mrs. Whitehead visiting rights.

In an interesting twist on the case, Mary Beth Whitehead (by then pregnant by another man and remarried) sued Mr. Keane, saying he had not adequately screened her for mental

and emotional barriers to carrying out the contract. In 1988, she won an out-of-court settlement reported at $30,000 to $40,000.

Why is "assisted reproduction" so popular now? In part, because it is there—it
is now possible to get help in reproduction, where it was not before, and child-
lessness seems less a judgment of God and more an obstacle to be overcome. But
more, infertility seems to be on the rise. Environmental factors could be involved
in infertility—pollution, especially chemical pollution of the water, has been suggested as a cause. More likely, the decisions on the part of a significant portion of
contemporary women to pursue careers and establish themselves in business and
professional life before seeking marriage and family have led to a general practice
of motherhood at an advanced age. The female body is most prepared for motherhood between the ages of 18 and 27; after that the viability of the eggs decreases
rapidly. We have become a culture where women do not marry young, and do not
wish to start families until they are sure they can provide for them as they want
(whether or not the children's fathers remain in the picture). Like all life choices,
this choice requires trade-offs, and easy pregnancy is one of them.

Pre-Implantation Genetic Screening: one reason to use In Vitro Fertilization
is that it permits Pre-implantation Genetic Diagnosis (PGD), screening for genetic
conditions present in the parents' families that they wish to avoid in the child. This
procedure involves teasing a single cell out of a blastocyst (in which all cells are
identical) and testing it for the suspected DNA. The success of this procedure,
and the fact that if no genetic disease was detected, the blastocyst would simply
supply the missing cell and go on to become a complete human being, led to the
possibility that we might be able to originate an embryonic stem-cell line without destroying an embryo, so avoiding the controversy outlined above. Objections
were promptly raised, that the pluripotent cell so detached could itself become a
full human being in the proper setting, so the controversy remains.

Genetic Enhancement: Testing for genetic disease always reveals the sex of
the child; should the procedure be used to get a desired sex? There are prospective
parents to whom the sex of the child is terribly important. This question raises the
further questions of the use of the IVF technology to promote genetic enhance-
ments. Right now we don't know what, if any, genetic patterns are associated with
the traits of above average height, blond hair, or musical ability, although we know
that they are genetic. The traits most desired by parents (or at least most under
discussion when the subject arises), like high intelligence and leadership qualities,
are surely highly complex in their genetic component, and more the product of
environment and early learning than of genetic endowment.

Note: the *technology* of Surrogate Motherhood, where the child is assigned to
his father, is the same as that of Artificial Insemination by Donor, where the child
is assigned to his birth mother. What differs is the understanding of the contracting
parties as to the ultimate home of the child.

Ought we to intervene to perform surgery on the developing fetus in the womb
to modify undesirable developments? What sorts of intervention might be appropriate? May we use surgical methods to enhance its life—intrauterine surgery—or
is that still too experimental to justify? If the mother decides that she cannot raise

this child, may we intervene surgically to abort the child? Or is the unborn child, at whatever stage of development, already a human person with full rights to life? This question has a tangled, and occasionally violent, history of its own.

1.4 The Abortion Controversy

Much of the debate has turned on whether or not the law should permit a woman to procure an abortion if she wanted one; that issue was settled by the Supreme Court in 1973, in the decision of *Roe v. Wade.*

> *Roe v. Wade*: First argued in December, 1971, *Roe v. Wade* (410 U.S. 113, 1973) was finally decided on January 22, 1973, and became the law of the land. To sum it up, the U. S. Supreme Court struck down as unconstitutional laws of Texas and Georgia that forbid all elective abortions except to save the life of the mother. The Court ruled that a selection of Liberty rights, derived from the fourth, fifth, and ninth amendments to the Constitution as an implied Right of Privacy, protected a woman's right to choose whether or not to bear a child. That ruling made abortion legal for the entire course of a pregnancy. In fact, only the first trimester abortion is protected by the right of privacy, as "between the woman and her physician;" in the second trimester, given the state's legitimate interest in the health of women, the state may require that abortions take place in legally regulated health care facilities; and in the third trimester, given the state's interest in the protection of life, the state may forbid abortions completely, except to save the life of the woman (or, in practice, when severe impairment has been discovered in the fetus). In the *very* extensive opinions, majority, concurring, and dissenting, virtually all questions on the permissibility of abortion are addressed—the implications for the health of women, the value of "potential" human life as opposed to actual human life, and the Constitutional protections extended to marital decisions, as found in the earlier decision on the use of contraceptives, *Griswold v. Connecticut* (381 U.S. 479, 1965)

Is this a "religious" question? Because of the early leadership of the Roman Catholic Church in public opposition to the legalization of the procedure, the abortion controversy has often been characterized as a "religious" controversy, between believers and non-believers. There is no basis to encourage or discourage the choice to abort fetus in the faith background of the United States; the Bible is silent on the issue. That has not stopped religious leaders from advocating for positions they feel are vital to their faith communities.

We can certainly acknowledge the value of human life without imputing moral "rights" that do not exist or challenging legal rights that do. How can this value be measured, and compared with other values? Is there any way of determining a generally accepted value of a developing life? Or is that value inevitably relative to social and political circumstances (the value of a male fetus in the womb of a queen feared to be barren, where the throne needs a male heir, for instance, versus the value of a badly impaired fetus who will surely be a drain on the society all of its life if it survives to birth)? It has been argued that the value of the fetus is relational, that is, as it is valued by the parents who will nurture its life and plan for its future. In such approaches, the question of the objective "value" of the fetus simply does not arise.

Meanwhile, the political battle goes on. The "pro-life" agenda includes lobbying for legislation that will forbid as many abortions as possible, given that they cannot get a universal prohibition. So in many states, such lobbies have encouraged legislation to require the parents' written consent before a minor can obtain an abortion. Courts have tended not to enforce such laws, as unjust to the pregnant girl; they are cognizant of the probability that in many cases of minors becoming pregnant, the minor's father or other close relative has fathered the child.

One success was a federal ban on the procedure known as intact dilation and extraction, or "partial birth abortion." In this procedure, performed late in the pregnancy or at any time that a dead fetus cannot be expelled normally, the fetal body is pulled from the mother's body by the feet ("breech extraction"), leaving the head inside the uterus; the physician then inserts a cannula into the base of the skull and evacuates enough of the contents of the head so that the rest of the fetus can be taken out, dead but intact. The procedure would be used when extreme hydrocephaly (the filling of the head with spinal fluid up to double its normal size) or other lethal condition made normal delivery impossible. The pro-life contingency took note of the fact that such hydrocephaly can develop without killing the fetus in utero, meaning that the physician actually ended the fetus' life by inserting the cannula, after part of its body was outside the woman's body, i.e. "born." And if the killing of a normal born baby is murder, then this procedure must be murder, or partial murder. Again, the hope was that if Congress could be induced to call *that* procedure impermissible, it would be a shorter step to calling all induced abortion impermissible.

In theory, intact dilation and extraction could be used to perform a very-late-term abortion on a healthy fetus, but no scenario has been proposed that would make sense of its use for that purpose. Legally, in most states, such abortions are illegal anyway (following the holding in *Roe v. Wade*, that states were permitted to pass laws forbidding third-trimester abortions save to save the woman's life.) If the fetus is that far along, (and the woman's life is in danger) abortion by hysterotomy, incising the uterus and physically removing the baby, as in a Caesarian section, would surely be used.

But don't fetuses have a right to life? The concept of "right to life" is spurious, essentially a rhetorical term born in politics, existing in no viable scheme of rights. We didn't have a "right" to be conceived and born, and we certainly have no "right" not to die, as we will all discover. In law and ethics, the notion of a "right" encompasses only those matters where exercise is a matter of choice. (There may be imputed rights for the incompetent—those rights which the incompetent would choose to exercise if competent—but these can generally be reduced to duties incumbent on the guardians.) If we insist on treating the abortion question as a matter of "rights," it turns out that it is the *same* right—the right not to have our bodies subjected to unconsented infringement—that is violated in the fetus if it is aborted and in the pregnant woman if she is forced to continue an undesired pregnancy.

There is certainly a legal right, on the part of a pregnant woman, to seek an abortion if that's what she wants. That legal right does not end the ethical discussion.

Whatever else it is, the termination of a pregnancy is always the ending of a human life, and is therefore always to be taken seriously; there can be better and worse reasons for electing to abort.

1.5 Dilemmas of Impaired Infants

When a child is born very badly impaired, either very premature or suffering from a genetic or other congenital handicap, what measures should be taken to preserve its life? Where no measures will succeed in saving the child (those born under 200 g, for instance), life-sustaining measures may clearly be withheld. Beyond that, this question does not permit of a general answer. But the dilemmas all turn on the value to the child of the impaired life that is foreseen for it, and the capacity of the parents to deal with it.

1.5.1 The History

The Johns Hopkins Baby: For most of human history, there was little that could be done for a child born with obvious deficits. The Greeks and Romans exposed children who were less than perfect, crippled or sickly, leaving them on hillsides to die of exposure (or possibly be adopted by people who wanted a child no matter how crippled). Through the first half of the 20th century, it was regarded as standard practice for an obstetrician to allow a clearly impaired child to die, telling the parents afterward that it was stillborn. (Note: until about 1960, anesthesia was liberally used during childbirth, so at the birth both mother and child were unconscious. The child would not start breathing unless stimulated, usually with a whack on the buttocks, so the physician had a few minutes to examine the status of the child, and decide whether it should live. The father was nowhere near the delivery room.) When a child was born with a severe impairment, especially one for which there was no cure, neonatal pediatricians would often allow the child to die painlessly in the newborn nursery (Dr. Ray Duff of Yale wrote compassionately of this practice). If the baby survived to go home, physicians often counseled the parents to put the child into an institution where it could be cared for until it inevitably died after a brief life.

In 1971, when the Johns Hopkins Baby was born, things were changing: there were now ways to resuscitate a child who was not breathing, some progress had been made on facilities for the handicapped, and most significantly, parents and child were awake and alert in the delivery room. Fathers now attended births, and little anesthesia was used. So when the Johns Hopkins Baby was born, the parents were very much part of the decision process. He was a male with Down's Syndrome, a trisomy on the 21st chromosome with implications of heart problems predicting a short life, severe mental retardation, and intestinal problems. His most pressing intestinal problem was duodenal atresia, a blockage of the intestines between the stomach and the small intestine. This condition is not compatible with life; no food can be processed until the atresia is surgically corrected. The parents had to choose whether or not to consent to the surgery; given the other implications of the Syndrome, they decided not to. So the infant was placed in a bassinet in the NICU, with a sign "Nothing by mouth," and left to die. It took almost two weeks for the infant to die (probably because the nurses were giving him water when no one was looking); publicized by a film made by the Hastings Center for Biomedical Ethics, the case aroused a small firestorm. Should the parents have consented to the surgery? Should the hospital and the state have allowed them to refuse? Does the baby have rights separate from the parents' judgment of his best interests? Did their decision amount to abuse?

The national debate occasioned by the Johns Hopkins Baby marked a changing scene in neonatal care: a new profession of pediatric intensivists was born, specializing in precisely those conditions that an earlier generation of pediatricians had pronounced untreatable, and an earlier generation of obstetricians had pronounced stillborn. The change in the attitude of the medical profession was echoed in popular attitude, as the next cases proved. By 1981 (in Danville, Illinois), when Pamela Mueller gave birth to conjoined twins, so seriously compromised that separation did not seem possible, some physicians counseled withholding treatment, which would not have been unusual in such cases two decades before, and with which the Muellers agreed. But this time other physicians disagreed, an anonymous caller alerted Protective Child Services, the state charged the Muellers with neglect and assumed custody of the twins. Several court battles later, the Muellers got their twins back. Surgeons attempted a complicated separation that killed one of the twins, but the other survived long enough to enter school. The controversy was still festering a year after the twins were born, when Infant (or Baby) Doe came along.

Infant Doe and the Baby Doe Regulations: Infant Doe was born on April 9, 1982, in Bloomington, Indiana, with Down's Syndrome and an intestinal defect, in this case a tracheoesophageal fistula, that made it impossible for him to eat normally. The question, as with the Johns Hopkins Baby, was whether to attempt surgery to repair the fistula or to allow the baby to die. Considering the anticipated difficulties of raising a Down's child, the parents elected not to consent to the surgery. The hospital's administrators disagreed, and asserted that there was a moral duty to treat the baby aggressively. They summoned a Monroe County judge, John Baker, who heard both sides out in a night meeting at the hospital, and concluded that the parents had every right to withhold consent to the surgery. (Of note, Infant Doe's father was a schoolteacher with long experience working with Down's Syndrome children.) The case would have ended there, but for the fact that an associate district attorney decided to take the case to the county circuit court, which reaffirmed Baker's decision, then to the Indiana Supreme Court, which did the same, and then to the United States Supreme Court (to Justice Paul Stevens, who had jurisdiction over that region). By this time the baby had died, but not before that district attorney had been on several news spots and talk shows affirming the child's "right to life," hence right to surgery. The publicity had two effects: it recruited the "pro-life movement," born in the wake of *Roe v. Wade,* into the cause of regulating the conduct of newborn nurseries, and it infuriated C. Everett Koop, then Ronald Reagan's Surgeon General, a pediatrician whose medical specialty had been surgery on infants. With Koop's encouragement, Reagan ordered the Departments of Justice and Health and Human Services (HHS) to mandate treatment in such cases, and make it a crime not to treat such infants.

States, not the federal government, have jurisdiction over such crimes as homicide and negligence, so the Justice Department was initially puzzled as to a legal route to enforce Reagan's will. Eventually it decided that cases of non-treatment of impaired newborns were cases of Discrimination against the Handicapped, and therefore in violation of the Rehabilitation Act of 1973 (section 504). Never mind that the law, an extension of the Civil Rights legislation, had been intended to apply to employment, public access and school settings, not medical settings, and to adults and children, not infants. HHS required a large poster to be put in all neonatal intensive care units, "Discriminatory Failure to Feed and Care for Handicapped Infants in This Facility is Prohibited by Federal Law." The poster was accompanied by a toll-free telephone number, the "Baby Doe Hotline," so anyone in an NICU could report "abuses" that they had observed—nurses, lawyers, parents, anyone at all. "Baby Doe Squads" of lawyers, physicians, and selected administrators would be dispatched to the location of any phone call to investigate, commandeer all medical records, and, if necessary, command physicians to perform treatments they would otherwise not have performed.

Misconceived from the beginning, the rules did not last long. In 1983, the American Academy of Pediatrics sued in federal court to block implementation of the rules, and won. Meanwhile, there had been some hotline reports, resulting in the deployment of Baby Doe squads; in no case did the squads find an endangered infant, but they surely had a mightily disruptive effect on the life of the hospital they visited. The final demise of the rules occurred in the *Baby Jane Doe* case.

Baby Jane Doe: The little girl who was known as "Baby Jane Doe" (Kerri Lynn, as her parents called her) was born on October 11, 1983, with *spina bifida* ("divided spine," a condition in which the spinal vertebrae have failed to close over and contain the central neurological structure, the spinal cord), and consequent *meningomyelocele* (or *myelomeningocele,* a condition in which the meninges and the nerve bundle of the spinal cord itself protrude through the opening in the vertebrae in a protrusion on the back of the baby). This condition almost always results in paralysis below the point where the meninges protrude, and a failure of the spinal fluid to drain properly through the spine, collecting instead in the brain in a condition known as hydrocephalus (water on the brain), which can crush the developing brain and result in severe mental retardation. Such a condition must be treated, if at all, with immediate surgery to close the wound in the back lest it become infected (meningitis), and the implantation of tubes called "shunts" to drain the liquid from the brain. Sometimes further surgery is necessary as the child grows; the prognosis varies, and as in the Baby Doe case, the physicians disagreed. Baby Jane Doe was transferred to Stony Brook Hospital on Long Island, because it had a better NICU. Dan and Linda, the parents, were presented with the choice: to authorize immediate multiple surgeries, or let the child die, with only comfort measures provided, death predicted within five days or so.

Badly conflicted (they had very much wanted the child), the parents decided against surgery, judging it to be "unkind" to the baby, opting for the comfort measures. Comfort measures included food, water, antibiotics, and "as much love as possible," provided by the parents. Kerri Lynn promptly defied predictions; her wound closed on its own, and she continued to live.

At this point a legal-political circus—there is no other word—began. Reporters, drawn by the publicity surrounding Baby Doe and the physicians' disagreement, brought the case to the newspapers. A Vermont Pro-Life municipal bonds lawyer, Lawrence Washburn, announced that he would rescue the baby, and filed suit to compel surgery. Several news conferences later, he was dispatched back to Vermont by Judge Melvyn Tannenbaum, who pointed out that he had no standing to sue in this matter. A local lawyer, William Weber, was appointed **Guardian *ad litem*** (the baby's legal guardian for this legal proceeding). Weber listened seriously to all sides as the situation changed beneath them—Kerri Lynn was not dying, but continued to recover. Eventually, even as lawsuits continued to drag through the courts, Weber authorized surgery to install shunts in Kerri Lynn's skull to drain the hydrocephalus. By now the Federal Government was alerted, and HHS announced that a Baby Doe Squad would shortly arrive to set matters right.

That was the end of the Baby Doe Rules. The Squad arrived and demanded the hospital records for Baby Jane Doe. The hospital adamantly refused to release the baby's medical records to the Squad. HHS appealed, and in 1984 the decision was made final—the Federal government had no right to seize and inspect hospital records. There was a further, snail-paced, appeal to the United States Supreme Court, which eventually ruled (*Bowen v. American Hospital Association et al.* 1986) that hospitals had no obligation to release records in such circumstances. The Baby Doe Regulations were gone.

Kerri Lynn went home with her parents in 1984, age 5½ months. Five years later she was tooling around in a wheelchair, talking, going to school, interacting with other children. She is still alive, but with serious deficits.

Lessons of the case of the Babies Doe: The Infant Doe case arose from converging streams in American life: First, the rapid advancement of medical technology and the competitive struggle among the hospitals to save more babies, at lower weights, with worse conditions; second, the increasing intrusiveness of the Pro-Life movement, that by now seemed to think it could parachute down anywhere and force matters to its preferred conclusions; and third, the contemporaneous assertiveness of the new "conservative" stream in American politics, epitomized by Reagan and Koop, that by now was poised for the long-term dominance of American politics. The Baby Jane Doe Case, on the other hand, reminds us that the American preference for privacy and the prerogative of the parents to make decisions for their babies has a deep consensus behind it.

The state of the question: controversies continue. Parents are generally concerned to make sure only the best is done for their baby, and inclined to ask that "everything" be done to save it. This concern seems to be more frequent and intense now, increased by the facts that mothers have babies much later, meaning that they have less chance to make it up if a first baby dies, also that babies born of compromised eggs of older years are more likely to have genetic problems, also that there are new areas of expertise in doing surgery on infants, opening tempting possibilities for medical and surgical experimentation. When parents demand surgery, or some other sort of exotic therapy to prolong the lives of those compromised by inevitably lethal conditions, Trisomy 13 and Trisomy 18, for instance—is it appropriate to encourage them?

May the irreparably impaired child—the anencephalic—be an organ donor to save the lives of other infants, even though a successful organ transplant will require the deliberate ending of the child's life?

The question of the acceptability of using an anencephalic child as an organ donor is beloved of philosophers, for it puts two of our central imperatives flat against each other. Of all impaired infants, the one most removed from the possibility of any really human life is the anencephalic baby: it has no brain. The cerebrum and cerebellum are absent, and the skull has not developed; only the top of the brainstem is visible as we look into its stunted uncovered head. Yet unlike other lethal conditions with which a child may be born, anencephaly does not afflict the rest of its organs—kidneys, liver, lungs and heart, intestines and skin—which can be healthy and normal at birth.

Anencephaly is a developmental problem, not genetic; it arises as an extreme case of spina bifida (Baby Jane Doe's impairment), where the neural tube does not close in the crucial early weeks of the pregnancy. The skull fails to develop, and the growth of the brain is stopped just above the brainstem. It is incompatible with life; the anencephalic generally dies in a matter of hours, days at the most. So its healthy organs are of no use to it at all. Meanwhile, babies are born every day with non-functioning organ systems, who might live a normal life if organs could be found for transplant. Unfortunately, it is much harder to find a suitable organ donor for an infant than for an adult, which is hard enough; most organ donors come from motorcycle and automobile accidents, and infants in such accidents rarely sustain the kind of injuries that would cause brain death yet leave the rest of the organs functioning, the only condition in which organs may be taken for donation. About 2,000 infants are born each year in need of a donated organ. Infants born with Potter's Syndrome (non-functioning kidneys) or hypoplastic left-heart syndrome (incomplete heart) are simply slated to die; yet if the anencephalic's organs could be transplanted into these infants (the surgery is not perfect, but the surgeons know how to ensure a very high probability of success), one anencephalic could save five or six other infants, who might go on to long and healthy lives. The case for donation of the anencephalic's organs is airtight, except for one disagreeable detail: the transplant surgeon has to kill the anencephalic on the operating table.

There is no other word for it. The infant has suffered no trauma that would render it brain dead, and it's very much warm and breathing. Of course we could wait for it to die, to lose all autonomic systems (heartbeat and breathing), but at that point the organs would be unusable—as the brain stem slowly dies, the organs lose oxygen, and by the time death is complete, they are irreversibly damaged. So we can't wait for that; we have to operate when the child is "brain absent" but not "brain dead." The surgeon has to place the infant on the operating table knowing that two or three of his planned incisions will end its short life.

Utilitarian reasoning yields the perfectly clear conclusion: we have here an infant with healthy organs who will be dead in a matter of hours—it is not being "deprived" of anything. It is not even a case of sacrificing one healthy infant in order to save several others (although, as John Rawls pointed out some years ago, Utilitarian reasoning could justify that, too); the anencephalic loses nothing. And if the surgery is performed, several infants who would have died will live full and flourishing lives. Add to the simple value of the organ recipients' lives the joy felt by their parents as their children are saved, and most especially add the joy of the mother of the anencephalic, who sees her tragedy turned into a miracle of saved lives and community rejoicing. How could there be any objection to the donation?

Yet deontological reasoning requires that we put the rule first, and ask, what would the world be like if such surgeries are authorized? Today it is the hopeless anencephalic whose life is ended to serve the needs of others. But nature is merciless to philosophers; every natural condition comes in degrees along a spectrum. If the skull starts to form too late in the life of the embryo, after the brain has already stopped growing at the brainstem level, we get hydranencephaly, where the brain is developed little more than the anencephalics' brain, but the skull is there, filled with liquid, so the child appears almost normal. Surely it would be no sacrifice to use hydranencephalics, too? There are other non-genetic developmental injuries—cerebral palsy comes to mind—that can result in total retardation, unawareness of surroundings, and generally a life that is barely sentient. These would be good candidates for organ donation, too. But then where do we draw the line? If we don't want to be asking at every point in the spectrum of impaired children, is it OK to chop *this* one up for parts—and we don't want to be asking that—we are much better off with a simple rule, that no child born and still even minimally alive shall be broken up for donations, no matter what joy would be occasioned by doing that.

Let's see how those arguments play out in a real case.

> ***Baby Theresa***: Laura Campo and Justin Pearson, an unmarried Fort Lauderdale couple, conceived a child in 1991. Because she had no health insurance, Laura did not seek prenatal care until she was eight months pregnant. That's when she found out that the child she was carrying was anencephalic, and it was too late to abort. She decided to carry the baby to term to donate its organs.
>
> After Theresa Ann Campo Pearson was born on March 21, 1992, the couple tried to arrange for organ donation, but the neonatologist said he would not harvest the organs unless she was declared brain dead (no neurological responses at all, flat EEG—the "Harvard" criteria adopted in all jurisdictions as a precondition for legal organ harvest: see below). Two judges told the couple that the organs could not be removed before brain death, and as the case was before the Florida Supreme Court, Theresa died.
>
> Laura Campo and Justin Pearson went on to advocate publicly for a change in Florida's laws to permit harvesting the organs of an anencephalic child, to no avail. The Supreme Court agreed to hear the case, even though moot, for purposes of determining policy, and after due deliberation, decided there should be no change in the law, which stands to this day in every US jurisdiction.

From the point of view of ethics, the Baby Theresa case is incomplete, for we never meet the projected beneficiaries—the infants whose lives would have been saved had the donation been allowed to go through. Surely these infants, and their families, were stakeholders in the issue; should they have been permitted a voice? (It is not clear that they were ever identified.)

1.6   Problems at the End of Life

How can a person exercise some control over the treatment he or she will receive as life ebbs to its close? Are the Living Will, the Advance Directives generally, the Health Care Agent, the Power of Attorney, the Conservator, adequate means to protect the patient's dignity to the end? How can we make sure that the patient's desires are put into effect? We might note at first, that the right to refuse treatment has not always been recognized.

Who should decide what medical treatment is necessary and what treatment should be withheld—the patient or the physician? For close onto 2,500 years, the answer was, the physician. The relationship between physician and patient was close to that of parent and child, or possibly priest and penitent. The physician had esoteric, highly complex, knowledge that he shared with no one (except other physicians of his school of medicine). It was never clear to the patient that the knowledge did not contain some special magic, or relationship to the gods, unfathomable by mortal standards. Physicians were taught, for those 2,500 years, to reassure patients no matter what the prognosis, not to discuss the patient's disease with him, lie to him if necessary to keep hopes up and compliance with the medical regimen strong, and if the patient balked, to enlist the family to keep the patient obedient to doctor's orders.

This attitude, "doctor knows best," has been characterized as baseless arrogance; actually, it had a solid scientific basis, possibly unknown to patient and physician alike. Beyond the treatment of obvious injuries—wounds, broken limbs—until about a century ago, the physician had very few remedies that actually worked well. (As a matter of fact, many commonly used medical treatments for internal ailments, treatments like bleeding and harsh laxatives, did more physical harm than good.) The physician rarely cured anything at all. Instead, he did what he could to allow the patient's body to heal itself, prescribing rest, warmth, bland food and soups, and occasionally giving opium or alcohol to dull pain. The physician's greatest contribution to the patient's welfare, in fact and in the theory taught in medical schools, was to relieve and prevent that stress that comes from fear of bodily illness and death, which was known to be harmful to the patient. Should the physician display lack of confidence, or hesitation, or deference to the patient's wishes, he would be stepping out of his role, and would no longer have the tonic effect on the patient's condition that they both had come to expect. That effect—the felt improvement in the patients' condition brought about simply by the confident reassurance and treatment of the physician into whose hands they had placed themselves—still exists; we call it the "placebo effect," from the Latin "I will please," and if we are doing clinical research (see Research unit, below) we have to control for it, or it will throw off our results. It is well to remember that for 2,500 years—until the twentieth century—the placebo effect was almost all we had.

In all those years, how did the physician deal with refusals—with patients who balked, said they did not want the treatment that the physician proposed? The physician generally categorized such refusals as what psychiatrists would call "denial"—the patient is so frightened, or deranged by his sickness, that he cannot see that what the doctor has ordered is really right for him. The fact that the patient demanded to have his wishes honored with regard to his treatment, and that the patient's wishes disagreed with the physician's judgment, was sufficient evidence to the physician that the patient was confused, panicked, or not in control of his nerves, and simply needed more reassurance, and possibly a little family coercion.

The stakes, and misunderstandings, escalate when the patient, or the patient's guardians, wish to refuse life-sustaining treatment, that is, treatment that will continue at least the heartbeat and provision of oxygen for the patient. If the patient refused such treatment, physicians tended to assume that the patient was "suicidal," and needed treatment with anti-depressants. If the guardians refused such treatment, the immediate conclusion was that they were trying to kill their ward (probably to get his money). Asserting the valid and controlling right in these cases, the right to have one's bodily integrity respected, was very difficult for patients and guardians alike. The issue of patient participation in the decision-making process surrounding medical treatment first entered the literature in a gruesome case of burn injuries: a young, active man terribly burned, trying to convince the doctors to let him die.

> **Dax Cowart**: Donald Cowart was 25, a bachelor who loved outdoor sports, and enjoyed working with his father, with whom he was very close. In April, 1973, he and his father drove their jeep to a parcel of land that his father was considering for purchase. Unknown to all, the valley was filled with propane gas, leaking from a buried pipe. When they returned to their car and started to leave, a spark from the jeep ignited the gas. A violent explosion and fire took the life of his father and left him with third-degree burns over two-thirds of his body. He should have died. But he was still alive when he reached the hospital, so the hospital kept him alive. The pain of massive third-degree burns is the worst we know; 45 years after Dax's accident, our burn units use sophisticated sedatives that keep their patients essentially asleep—they will never remember their time in the burn unit. Dax was not so lucky. The pain was excruciating, the physicians were afraid to give him adequate pain relief for fear of killing him, and they insisted on moving him from his bed and bathing him in chlorinated disinfectant twice a day for fear of infection. (That made the pain worse, much worse.)
>
> Dax argued that he had nothing left to give him any quality of life. He was blind and almost deaf (that, interestingly, is how he acquired the nickname "Dax": the physicians discovered that he could hear "Dax" when he could not hear "Donald." He later adopted the name as his own.) His hands were burned so grotesquely that he would never be able to use them properly again; his legs were badly disabled (he could crawl). He had lost his father and partner, he could not work, and he could never again take part in all the activities he had enjoyed. He pleaded with the physicians to leave him alone and let him die; they refused, even though a psychiatric evaluation had shown Dax to be competent to make his own decisions. After years of treatments, changes of facilities, at least three suicide attempts followed by reluctant agreements to resume treatment, Dax started functioning again—won a lawsuit against the gas company to give him some income, completed law school, practiced law successfully for awhile, married, and settled into a reasonably comfortable life. He argues to this day that his wishes should have been respected, and that he should have been allowed to die as soon as the extent of his injuries was known.

Dax's case was the first. In 1973 and for all the years prior, no one would have thought of respecting the wishes of a person who refused life-sustaining treatment. But Dax's rational single-mindedness raised the question: whose life is it, anyway? Why should Dax be forced to accept excruciating treatment against his will? Should Dax have been accorded the right to make his own decisions? The case was written up for the bioethics literature; as with the Johns Hopkins baby, a short film was made and circulated; there was a much-noticed Broadway play, *Whose Life Is It, Anyway*? The right to refuse treatment was not established by Dax's case, but he certainly put the question on the agenda of bioethics.

The case that did establish that right, and brought the questions to the general public in unforgettable form, arose in the worst circumstances that could be imagined—a decision to provide or withhold life-sustaining treatment of a beautiful young woman caught in the middle of a medical controversy that ended in the New Jersey Supreme Court.

> **Karen Ann Quinlan**: Karen Quinlan was 21 when she fell into a coma during a party; she apparently had not eaten that day, had taken barbiturates of some sort, and consumed

alcohol, which intensifies the effects of barbiturates, especially on an empty stomach. She retired to a bedroom in the house to take a nap. When she was found, she was not breathing; no one knew how long she had been that way. They rushed her to St. Clare's Hospital in Denville, NJ, and put her on a ventilator (or respirator) to restore lung function. This was in April, 1975. She was kept on the ventilator all summer, making no progress toward regaining consciousness. In the fall, her parents reluctantly concluded that she would not regain consciousness, and asked the hospital to remove the ventilator.

Unsurprisingly, the hospital refused; the AMA had declared all removal of life-support to be "euthanasia," mercy-killing, which amounts (if the patient's opinion does not count, and it did not) to murder, so the physicians were understandably unwilling to go along. The hospital was afraid that if they acceded to the family's wishes, the family would change its mind and sue for malpractice. (Recall, "malpractice" can be shown to have occurred if (1) the patient outcome was undesirable, and death is usually held to be undesirable, and (2) the physician's behavior is a "departure from normal standards of medical practice in a community," and taking a patient off a ventilator to die, in 1975, was just that.) There followed a battle of epic proportions, all over the newspapers. The Quinlans hired a lawyer, Paul Armstrong, who offered to sue to have Joseph Quinlan appointed Karen's legal guardian (she was an adult, but she was now incompetent, so a guardian was needed) for purposes of allowing Karen to die; the argument turned on his contention that Karen was being denied her civil rights, specifically the "right to die." The theory was unworkable; there is no "right to die," anymore than a "right to life," for most of the same reasons. There is a right to refuse treatment, but that right had yet to be established in the courts; remember this was 1975. A simple request to have her father appointed as her guardian might have succeeded; the assertion of a "right to die" did not.

The lawyers came back with a petition to have her father appointed guardian; that was done; Joseph Quinlan requested (ordered, now that he was court-appointed) the removal of the ventilator, so the hospital did it—very slowly, over a period of weeks, slowly enough so that Karen's young body could be weaned from the ventilator and start breathing on its own. Now what? The physicians who had examined Karen were clear that her brain was gone, had most likely been gone before she was found and rushed to the hospital. The brain cannot survive without oxygen, and deprived of it, it is the first part of the body to die.

Karen was in a coma, in the kind of coma that we now call a "persistent vegetative state (PVS)" or "minimally conscious state." In PVS, the patient does not just lie there in a deep sleep. There are periods of waking and sleeping, tossing, sometimes thrashing around at first, making noises, mostly moans of some sort, perspiring, opening and closing the eyes. Further, from inactivity and disconnect between the brain and the muscles, Karen had become rigid with contractures, when the body bends itself back into a fetal position just because there are no muscles to keep the limbs extended. It was not a pretty sight, and it would not end for ten years. Karen was transferred to a nursing home (there was nothing more that the hospital could do for her, or to her), and lay in PVS until June 1986. Eventually pneumonia, "the old man's friend," claimed her life.

The Quinlan case was a true landmark; let us review its significance.

Until *Matter of Quinlan,* medicine, law, religion and the public at large were in agreement: death is the worst possible outcome of any medical situation; when a patient is very sick, physicians will do *everything* to preserve the patient's life, however minimal that life may be, in the course of trying to restore all biological functioning and getting the patient well. This shared conviction had lasted for 2,500 years. But until the 1970s, there had been few ways to prolong the life of the dying patient. One organ system or another would fail, and there was nothing to be done for that, then the heart and lungs would fail, and the old ways of determining death—put a mirror in front of the mouth to see if the patient is breathing, put your ear on the chest to listen for a heartbeat—would tell you soon enough when the battle for life was over. (Dax Cowart would have been dead of infections within days of arriving at the hospital.) At the end, only opiates would make a difference (and, schooled in the understanding that it is appropriate and expected for death to be accompanied by pain, they didn't use very many of those).

The 1960s and 1970s had seen revolutionary advances in mechanical means of substituting for failing bodies—dialysis machines, ventilators, defibrillators and mechanical massage to restart stopped hearts, nasogastric and intravenous feeding tubes—and had changed the medical landscape forever. From that point on, patients could be kept minimally alive for indefinite periods after the brain was gone and the human had disappeared. Two separate (and very distinct) questions were brought to public debate by these developments, and they tore the field of bioethics apart—even as they created it. Until this point, physicians and other practitioners had had no idea there were serious philosophical questions lurking at the base of their practice, and until this point, philosophers had had no interest in exploring the philosophic bases of professional practice in any field, including their own. Now suddenly there were two deeply theoretical questions in dispute, important—literally, matters of life and death—and both medicine and philosophy had to learn how to solve them. For starters, they had to learn how to think about them.

The first question raised was, what is death? When is a person dead? Our understandings of the determination of death in the mid 1960s were not far from those that had accompanied the human race from its beginnings—see if the person is breathing, listen for a heartbeat, and if those are not there, the person is dead, and the physician would pronounce death. (We now call this "cardio-pulmonary death.") That definition had been rendered obsolete in the early 1960s by the development of the "heart-lung machine" which could take over breathing and heart functioning for a patient during open-heart surgery—but no one noticed the impact on the definition of death, since the heart–lung machine was only used for short periods of time in the process of trying to heal someone. Then, in 1967, Denise Darvall suffered irreversible brain damage in an automobile accident in South Africa, and Dr. Christiaan Barnard transplanted her still-living heart, kept alive on a heart–lung machine, into a dying cardiac patient, Louis Washkansky. But if Denise's heart was still alive—and that was the point of the transplant, after all—she couldn't possibly have been dead by cardio-pulmonary standards. Was she really dead?

The question could not be ignored, because organ transplantation held out new and vital hope for patients dying of organ failure. To retrieve viable hearts and livers from cadavers, we have to be able to keep those organs oxygenated, and that means we have to keep oxygenated blood flowing through them, and that means we have to keep the heart pumping and the lungs breathing. (Kidneys had been transplanted, especially between twins, for some time. Kidneys don't count, because we can take a kidney from a healthy person without hurting him.) But for most organ transplantation, we have to know that the donor is really dead—we certainly don't want to be harvesting organs from people who are still alive. We had to balance our fear of violating the physical integrity of a living human with the immeasurable benefit of saving a life, or many lives, by harvesting and transplanting healthy organs—we had to find a way to declare death while the machines were still operating. Harvard Medical School, after much debate, came up with criteria for "brain death" (called at that time "irreversible coma") that would satisfy these conditions. The patient had to show no spontaneous movement or breathing (meaning that you had to take the patient off the ventilator very briefly, to see if he would try to breathe), no response to physical stimuli, and two isoelectric (nearly flat) electroencephalogram (EEG) readings 24 h apart. Then the patient was dead by the Harvard criteria, and his organs could be transplanted into patients who needed them.

The definition was still controversial. There were jurisdictions that simply did not accept the Harvard criteria, and in those jurisdictions no organ transplants could be performed. Other writers argued that the "whole brain death" criterion demanded by Harvard (which included brain stem death—since spontaneous breathing defeated the criteria) limited organ transplantation unduly. It seemed to them that when the person was irreversibly lost to human life and communication, which happened with "cerebral death" (the death of all the higher parts of the brain), death should be pronounced and organ transplantation permitted. That would define all patients in PVS as "dead," and make them eligible for transplant. The argument was not successful—whatever else Karen Ann Quinlan was for those 10 years in the nursing home, blinking, perspiring, making noises, she certainly did not seem to be "dead." (After all, if organ transplantation was not contemplated, and she was "dead," we should call the funeral home; the notion of having her funeral, and burying her, in that condition, was beyond repulsive.) This consensus extends until today; on these grounds in the 1990s, recall, it was found to be unacceptable to donate the organs of an anencephalic baby, Baby Theresa.

The second question raised was, when is it permissible to end life support and let a person die? The 1950s answer to the question, unchanged since Hippocrates, was Never. But now a person couldn't even die in peace; now we had machines that could re-start the heart and machines that could "breathe" for the person through tubes down into the lungs. Now when a patient died in the hospital (i.e. the patient stopped breathing and the heart stopped beating), a well-trained staff would arrive with a "crash cart" to resuscitate and intubate the patient—to start the heart beating again (usually with electric shock) and pump oxygen into the lungs. The patient was in no shape at that point to refuse that treatment, and usually there was no one else around to speak for the patient—even if anyone had been willing to listen. Yet it seemed clearly wrong from every point of view to continue these resuscitations on a patient who would never recover even to leave the hospital. In many cases the patient did not want treatment continued, the family wanted only to let the patient die with some semblance of dignity, the resuscitations were clearly futile, in that they would not accomplish any reasonable goals of treatment, and they were consuming hospital resources for nothing.

In short, these coerced treatments, these resuscitations, would be ruled out by any considerations of patient autonomy, and so violated the imperative of respect for moral agency, or respect for persons; they violated the imperative of benefi-cence, for they not only did not produce happiness for the greatest number, they produced misery all around; and they violated the most basic considerations of justice. Yet they were held to be "morally required" by physicians and nurses trained in the 1950s or before, on the basis only of an imperative to prolong life at all costs—and that tradition was getting a very hard second look.

It was fashionable to blame the quagmire on religion. Even at the time of the Quinlan case, editorial voices were heard to the effect that the "Catholic Church" taught that life had to be prolonged indefinitely, and that "the Church" had required that St. Clare's hospital, a Catholic hospital, refuse removal of the ven-tilator. That claim happened to be completely wrong (as the Quinlans' Catholic parish priest knew; he supported the Quinlans from the beginning in their efforts to have Karen removed from life-support.)

The Roman Catholic Church had issued the Declaration on Euthanasia, addressing the problems and obligations of death and dying, in 1961 (it's been updated several times since). What it said, clearly enough, was that it was never justified to kill a patient, to take an action aimed solely and effectively and end-ing his life, any more than it was justified to kill any innocent person. Where there was a prospect of recovery, patients and physicians alike were under an obliga-tion to work to get the patient well, back to work and family. On the other hand, if it was clear that the patient was dying, the patient was under no obligation to accept further treatment, nor was the physician under an obligation to offer it. It was assumed that all "ordinary" means of life support would continue—offering food and water, keeping the patient warm, clean, presentable and free of pain and anxiety—but no "extraordinary" means were required. (In later versions, all "pro-portionate" means were required, i.e. where the expected benefit of the treatment compared well to the burden imposed, but "disproportionate" means, where the burdens of the treatment outweighed the expected benefits, might be foregone.) The Church was alone, at that time, in considering *all* the burdens that the phy-sician's latest experimental therapies might impose on the dying patient and his family: not just pain, nausea, anxiety, and all the other discomforts of medical interventions, but also financial burdens, the burdens of travel if the treatment had to be carried out at a distant center, and the loss of time from work and family incurred by the caregivers. A whole, rounded, view of human life, death and fam-ily is contained in that document, and the Catholic Church represented at once the most reasonable (in terms of utility and justice) position on the contentious spectrum of positions on death, and at the same time the most spiritual position, respectful of the natural God-given limits of the earthly life of humans and of the natural concerns of patient and family. In the consensus that began to emerge in the late 1970s and 1980s, on the moral obligations surrounding death and dying, Roman Catholic theologians, notably John Paris SJ and Daniel Maguire, were sig-nificant contributors.

The court decisions in the Quinlan case had more impact on hospital proce-dures than on the ethical debate. Clearly the New Jersey Supreme Court was con-flicted on its decision; never before had a court had to decide that it was OK, sort of, for a person to die. So it recommended a procedure for hospitals to follow in all future cases: convene an Ethics Committee to track and give opinions on these cases, discussing the prognosis for the patient, the needs of all the stakeholders, and the best solution for patient and hospital. Since then most hospitals have maintained Ethics Committees, meeting once a month or so to discuss the ethical dimensions of the problems faced by the hospital. The most frequent problem encountered is the first one encountered; when the biological functioning of a dying patient is being maintained on high-technology life support, at what point is it permissible, recommended, even obligatory, to remove the life support and allow the patient to die?

The key question in the Quinlan case was about the removal of the ventilator. When the ventilator was removed, and (as a result of careful control of the process by the hospital staff) Karen kept on breathing, the Quinlans (informed by the nursing home that feeding was part of "ordinary care") did not consider the possibility of requesting the withdrawal of the feeding tube. It was not until 10 years later, when pneumonia took her life, that Karen died. The case of Nancy Cruzan raised the question of the feeding tube in interesting ways, with interesting results.

Nancy Cruzan: Nancy Cruzan's car skidded off an icy road in Missouri, on January 11, 1983. She was thrown from the car into a watery ditch. When the paramedics arrived, they found that she was not breathing, and injected a stimulant into her heart to bring her back to life. They succeeded, in a way that has given many of us pause: she survived, but never woke up, never communicated with anyone ever again, lay there in contractures for seven years, breathing but fed with a feeding tube, while her parents grieved and the state of Missouri paid $130,000 per year for her care. At the end of seven years, Joe and Joyce Cruzan asked that the feeding tube be removed. The probate court granted their request, but the Missouri Supreme Court, citing a state interest in protecting the lives of incompetents, overruled the court.

The Missouri Court said, interestingly, that it would have granted the request had they had enough evidence that Nancy herself had expressed a wish that she not be maintained in this state. There are several standards of evidence in the law: a preponderance of evidence means that there is more evidence for one conclusion than evidence against it; clear and convincing evidence, a more rigorous standard, looks for papers and signatures and signs of deliberate acts; evidence beyond a reasonable doubt is the extreme case, used when the question is one of convicting an accused of a capital crime carrying the death penalty. The parents' testimony that Nancy had expressed a disinclination to live in these circumstances, and their opinion, as her guardians, that it was in her best interest that the treatments be ended, constituted a preponderance of evidence, the Court agreed, but they wanted a higher standard, clear and convincing evidence, which would be supplied, for instance by a Living Will, or Advance Directive, signed by Nancy while she lived, specifying what medical interventions she would want refused on her behalf should she be in a terminal (irreversible) condition and become incompetent to make refusals for herself.

"Living Wills," or "Advance Directives," had been in the literature for a long time, praised by ethicists because they allowed people to exercise autonomy even when they became unable to make decisions for themselves, disliked by physicians (at first) because they limited the options open to the medical care providers, disliked by others who do not approve of any permission to end life support. The Living Will has no force until two conditions have been met: a physician must have declared the patient to be in a "terminal" condition (he's not going to get any better), and the patient must have lapsed into some kind of coma, and cannot any longer make decisions for himself. A typical Living Will specifies a patient's preference that, should his heart and breathing stop, he should not be resuscitated to start the heart and he should not be intubated to restore breathing. It continues more generally to say that any life-sustaining interventions should be stopped at that time—dialysis, stimulants, antibiotics, whatever, and any artificially provided nutrition and hydration—like feeding tubes.

The hospital will translate the Living Will into Medical Orders, entered on the medical chart: DNR (do not resuscitate), DNI (do not intubate), No ANH (artificial nutrition and hydration), CMO (comfort measures only). When these orders are in effect, certain kinds of treatment/care will continue to be provided: the patient will be kept clean, warm, free of pain and anxiety to the best of everyone's ability, and his privacy will be respected.

The Cruzans appealed the case to the U.S. Supreme Court, which decided (in July 1990) that Missouri did indeed have a right to demand a higher standard of evidence for cases involving life and death. So the Cruzans lost. But three much more important holdings were incorporated in that decision: (1) For the first time the U.S. Supreme Court recognized the Constitutional right (under the doctrine of Privacy, see *Griswold v. Connecticut*) of a competent patient to refuse unwanted treatment (that had been accepted for some time, articulated by the N.J. Supreme Court in the Quinlan case, but had never before been articulated by the U.S. Supreme Court); (2) accepting the argument of the American Medical Association in its *amicus curiae* brief, the Court found that taking out a feeding tube was the same as withdrawing or withholding any other kind of life-sustaining intervention. Sentimental editorials likening the provision of feeding tubes for the terminal patient to bottle-feeding infants, arguing that feeding was so central to the human experience that stopping it constituted moral abandonment, did not carry the day on this issue; (3) in a concurring opinion, Sandra Day O'Connor emphasized and underscored the desirability of Living Wills. Had there been a piece of paper, she noted, setting forth Nancy's own desires, the Missouri Court might very well have ruled the other way.

Meanwhile, the major holding was that states have a perfect right to adopt stricter standards of evidence for these cases. The Cruzans went back to Missouri, looked for more people that might have talked to Nancy about such matters, found that many of her friends knew her under her (briefly) married name of Davis, and these friends came forward to testify to conversations that many people, of all ages, now have, about the end of their days. A hearing that included that testimony determined that the Cruzans had met the "clear and convincing" standard, and on December 14, 1990, the feeding tube was removed and the rest of Nancy died.

In response to the Cruzan decision, the U.S Congress took up the question of Living Wills in its last session of 1990 (called OBRA, the "Omnibus Budget Reconciliation Act"). Into that Act many strange laws can be folded, and one that showed up in 1990 was the "Patient Self Determination Act," PSDA. In that law, in-patient care providers are required to find out from every admitted patient, on admittance, whether or not they have a living will. If so, the patient is asked to produce it; if not, the patient is offered help in executing one. This law has inserted the Living Will into the middle of the health care system, raising expectations that each person, eventually, will be able to make his own decisions about how he wants to die.

By this time, it might have been said, we had a deal, a three part consensus. Death was not the worst thing that could happen; much worse was "living" on and on, with no consciousness of anything around, dying expensively prolonged with elaborate machines. (Most of the American public will agree with this.) So even if death loomed, you had a Constitutional right of privacy to refuse those interventions that promised to extend life, at whatever minimal level. Further, after Quinlan, your surrogate decision makers—conservators, guardians, health care agents, depending on the jurisdiction and circumstances of appointment—can make that decision for you. The right of privacy is not lost when you become incompetent. So you have a right to make a Living Will, and have it enforced by a person you appoint to be your surrogate. A third element of the emerging consensus was fragile as we turned to the new millennium: when the patient is terminal, and incompetent, and when no measures are likely to do the patient any good, the hospital may refuse the pleas of emotional relatives to continue treatment, any treatment, for the sake of preserving "hope"; this refusal is made on grounds of "futility." Hospitals have never dealt well with emotional patients and especially with emotional parents or children of an incompetent moribund patient. Most of the difficulties came to roost in the case of Terri Schiavo.

Terri Schiavo: On February 25, 1990, 27-year old Terri Schindler Schiavo collapsed, stopped breathing, and by the time medical help could be procured, had suffered brain damage from anoxia (loss of oxygen). The similarities to the Quinlan case are striking: it was alleged (not proved) that over-dieting, even anorexia, had something to do with the collapse, and other lifestyle factors were brought into the account; in fact, no one knows to this day, in either case, the nature of the insult to the central nervous system that resulted in the comatose state for either woman; in both cases, the guardians' efforts at finding some way to aid the recovery of their ward was futile; in both cases, when the guardians finally realized that nothing would do any good, and asked for suspension of the life-support that had been intended to buy time until a cure was found, they met with resistance from outside parties, often declared in the name of religious belief. In Terri's case, the resistance was a media firestorm.

Two months after Terri's collapse, her husband Michael (the legally designated guardian) placed her in a rehabilitation center; then in an experimental "thalamic stimulation" center in California; aides were hired, and worked for two years trying to get Terri back to herself. Terri was breathing on her own, but required a percutaneous endoscopic gastronomy (PEG) feeding tube; unlike the nasogastric tube that had fed Karen Quinlan, Terri required a surgically implanted tube to take nutrition directly into her stomach. Terri's parents, Robert and Mary Schindler, actively involved in Terri's care, at one point tried to take care of her at their home, but were overwhelmed by the task and sent her back to Michael. From 1991 to 1994, Terri was in Sable Palms, a skilled nursing center where several levels of therapists worked on restoring function. At this point Michael received a sizeable settlement from a malpractice suit, to be devoted to Terri's care. The Schindlers wanted part of the settlement, and apparently the disagreement over that allotment precipitated later hostilities. In the next four years, while Terri's condition did not change, Michael became a licensed respiratory therapist, which would make him eligible to oversee more of Terri's care.

In May 1998, convinced by now that Terri was not going to get better, Michael asked the probate court in his Florida district to permit removal of the PEG. The Schindlers opposed the request; two years later, the court agreed with Michael. Over the next year and a half, the Schindlers appealed the ruling to two higher courts, losing each time. The Schindlers turned to the press, which cooperatively published enough pictures and stories of Terri to get Florida Governor Jeb Bush to take the side of the parents, U.S. President George W. Bush to praise the Schindlers for their defense of their daughter, Senate Majority leader Dr. William Frist to declare, on the basis of the videotapes of Terri, that she was probably not in a coma, and an uncounted number of swamis, monks, experimental therapists hawking their coma-ending panaceas, religious fund-raising organizations, media hounds of every sort, drawn to the Florida bedside of one unconscious woman, to argue about how her guardian ought to resolve her problems. There were accusations that Michael Schiavo had abused Terri, and that the abuse was the cause of the heart attack she had suffered, and that therefore he should not be guardian. There were Congressional Hearings and prayer vigils, President Bush flying in from vacation to sign "Terri's Law" which purported to have the authority to make everyone keep Terri's PEG in place. Enormous pressure was put on Jeb Bush to send in the Florida Militia (illegally) to place a guard over Terri so the PEG could not be removed; thankfully, he refused to do that. The media circus continued through March 2005, when the FBI arrested a man who had offered a quarter million dollar bounty for the life of Michael Schiavo.

The courts continued to do their job. Three times the U.S. Supreme Court refused to intervene in the case, saying the lower courts had done their job well. In the middle of March, 2005, the PEG was removed; 13 days later Terri's body died. Autopsy showed no spousal abuse, and, surprisingly, no heart attack; it also showed massive brain damage, incurred at the time of the collapse, rendering all therapeutic attempts futile from the outset.

This case was very troubling for the bioethical community. After all, we thought we had a deal. When an incompetent patient's legal spokesman, having tried all remedies that might hold out hope for recovery, concludes that the battle is lost and asks for the end of life-sustaining interventions, the interventions are supposed to stop. *Quinlan* and *Cruzan* determined that. Now all of a sudden there is a major challenge to the consensus that had been worked out in the 1970s and 1980s. Now, suddenly, all is political. Right-wing forces claiming that Terri was being "murdered" grabbed the headlines, interfered with the patient's care, violated the privacy of everyone involved, and turned a private family tragedy into a three-ring circus, with the peddlers of miraculous cures in the north ring, the prayer vigils in the south ring, and in the center ring, the President of the United States, leaping from Air Force One to sign into law a bill ordering medical treatment for one lady in PVS, unconscious of the whole commotion ordered in her name, unable to benefit from any of the efforts of her would-be "supporters." We found out quickly enough the motive behind the circus: across the country, organizations backing conservative political candidates had been speaking to church groups about "Poor Terri," and collecting money for "Terri's legal and medical expenses," which needed no subsidies at all. They were honest enough; when asked what would happen if they collected more money than was needed for Terri, they said the money would go to "other conservative causes." That it did, no doubt.

The Schiavo affair led to general revulsion at such media spectaculars. When the dust settled, the consensus was back in place, President Bush's approval ratings sank another several points, and Dr. William Frist's political career ambitions were brought to a rude halt. The American people really don't like these family matters splattered all over the highways.

Not all questions are resolved even when the consensus prevails. Terri Schiavo at least had a clearly recognized guardian to make decisions. What if there is none? In the absence of advance directives or a designated surrogate decision-maker, how shall we decide when further medical interventions are not indicated? How do we decide when to write Do Not Resuscitate Orders? When is it clear that the patient should be placed on Comfort Measures Only (no further treatments, ventilation, nutrition, hydration, tests or monitoring)? Sometimes there are conflicts: What do we do when the patient is terminal and incompetent, there are no advance directives, and the family is in violent disagreement about what to do?

Relief of pain: The medical profession's orientation toward pain relief presents us with an interesting case of change conditioned on scientific findings. The prevailing view until the mid-1980s was that the use of morphine and other opiates shortens the life of the terminal patient, by depressing respiration. The moral question for the profession was, is it permissible to administer opiates to relieve a terminal patient's unbearable pain, when we know that eventually the dose will end the patient's life? Or, which is more important, the quantity (length) of life or its quality (pain-free)? A consensus had emerged that it was more important to relieve the pain than to extend life the last possible minute: better to die pain-free on Tuesday than screaming in agony on Wednesday. Studies during the 1980s, however, raised the intriguing possibility that morphine might actually extend the life; that pain itself placed the body in such stress that survival was much shorter than it might be with adequate pain relief. These findings have greatly encouraged physicians, nurses and other health care personnel to attend without hesitation to the relief of pain, without worrying about "hastening death." (Patients, who worry about dying in pain much more than they worry about dying, are greatly encouraged too.)

"Futility": Let us review: Prior to the 1970s, the physician decided what treatment was appropriate for every patient, and ordered it, persuading the patient and family to agree if possible. The court cases of the 1970s–1980s decided that patients (or their surrogates, if the patients were incapacitated) could refuse treatment, even life-sustaining treatment, at the point of the intervention or in advance, through a Living Will, and that their decision prevailed even if the medical staff judged otherwise. Now, what if a terminally comatose patient is beyond all medical help, but can survive for an indefinitely long period on life support, and the medical staff judges that it is time to end life support, but the family wants life support continued—indefinitely? Can the hospital, citing respect for the dignity of the patient (it is no one's desire to be kept "alive," unconscious and unable to communicate, while the body starts to decompose), insist on pulling the plug? Or can it cite the unjustified use of scarce hospital resources? Or must it accede to the family's desires?

It seemed to the bioethicists that when families, for a variety of reasons, decided against medical advice to insist on the continuation of ventilators and feeding tubes, hospitals should advance the notion of "futility" to justify an unilateral decision on the hospital's part to withdraw life support—that is, that if it was agreed that continuing life support interventions for this patient was "futile," not conducive to the achievement of any worthwhile medical outcome, then the hospital should be able to order the removal of life support whether or not the patient's family agreed. Why did hospitals want to do this? First, they did not always find the family's reasons compelling (in one case in my experience, the family wished the patient kept alive because they were dependent on his Social Security payments; in another, the closest family member was living rent-free in the patient's house, which would have to be sold for the division of the inheritance should he die; in a third, the patient had won one of several enormously rich Connecticut lotteries and elected lifetime payout, which would stop if he died). Second, they regarded the occupation of an Intensive Care Unit bed, and 24-hr use of scarce machines whose operations had to be monitored by hospital staff needed for other patients, an unjust demand on resources and (on their own part) bad stewardship of hospital capacity.

Then should every hospital have a "futility" policy, saying that when medical staff conclude that there are no further interventions which can benefit the patient in any meaningful way, or accomplish any goals beyond the continuation of heart and lung function, then at that point further hospital treatment should be called "futile," and stopped, whatever the family might say? Such a policy would be problematic; this is one of the points in the Ethics of Health Care where legal actions and fears intrude.

As above, in the 1970s and 1980s the courts gradually acquiesced to the right of the patient to refuse treatment even against medical advice, citing the patient's right to control access to his own body, and the patient's autonomy. Then the families of dying patients, who had been advised that it was really time to withdraw life support, went back to court to assert the right to demand continuation of life support even against medical advice. The probate courts, to the distress of medical staff and hospitals, generally agreed. After all, if the patient or surrogate is the decision-maker in cases where the staff wants life-support continued but the patient or surrogate does not, it would follow that the patient or surrogate is the decision-maker when the case goes the other way, right? Or so the courts reasoned.

Bioethicists flatly disagreed. The way they saw it, the physician (health care staff) had the duty to offer "only those measures that are for the benefit of the patient," as one translation of the Hippocratic Oath has it; or, "I will do nothing that is not for the benefit of the patient," as another version puts it. If a medicine, or a treatment, will not benefit a patient, the physician has no right to offer it; if an ongoing regimen or treatment, even a life-support treatment, has ceased to benefit the patient, or it becomes clear that it will not benefit the patient, the physician is obligated to withdraw it. The duty not to offer, or continue, treatment that is not doing any good, "futile" treatment, needs no new law or policy; it's in the Hippocratic Oath. Bioethicists saw the matter as one of "paired vetoes": the physician could, indeed was obligated to, refuse to offer any treatment that would confer no benefit on the patient, and the patient could refuse any treatment that was offered. That's not the way several courts have seen it; all the judges want to know is, who has the right to make the decision in this case? And if ethicists insisted that the patient had that right in cases of refusal of treatment, then they must have the same right in the case of demanding treatment.

Physicians and hospitals are distinctly leery of Futility Policies, as an open invitation to litigation. Families who opposed the withdrawal of life-support interventions are likely to sue physician and hospital alike for "wrongful death" if they are in fact withdrawn, even if the patient had a Living Will insisting that life support be withdrawn under the circumstances that obtained. The law requires that if the Living Will, executed when the patient was competent, contradicts the family's wishes, the Living Will shall prevail. (If the patient reasonably foresees such opposition from the family, he is free to appoint a non-family member as surrogate decision-maker, or "health care agent." In the early days of AIDS, 1980s and 1990s, AIDS patients were well advised to do that.) But as hospitals knew at the time, dead patients don't sue, but their families do: if there is strong opposition to carrying out the terms of the Living Will, the hospital will not insist on them. If the hospital that withdraws life support has a Futility Policy in place, and the whole matter ends up in court, that policy, whether or not it was invoked in this case, will be cited by the family as evidence of "intent to abandon" patients that the hospital no longer wished to continue on life support (probably, it would be darkly suggested, because they were costing too much money).

Most hospitals do not have clear Futility Policies. Instead, they try with all their might to find some skilled nursing facility, convalescent home, any Long Term Care facility, which will take the patient on terms the family can agree to. That gets the problem off the premises. If that does not work, nor several family conferences, counseling from the social worker and the chaplain, meeting with the ethics committee—the hospital usually just puts up with keeping the patient on life support until death.

1.7 Voluntary Death and Assisted Suicide

As the Quinlans found out in their misdirected approach to the New Jersey courts, there is no "right to die." Eventually a "right to refuse treatment" was established; that includes the right to refuse ANH (Artificial Nutrition and Hydration), the provision of food and water by tubes, however administered. Can a patient refuse ordinary food, and elect to starve himself to death? It would seem so; eating, swallowing, in a patient not crippled by brain damage, is a voluntary act which the patient can refuse to do. Unless the patient is declared to be incompetent, unable to make decisions for himself, no one will arrive with a feeding tube and force him to eat. Can the patient require that the hospital assist him to starve to death, by giving him support and painkillers to take away the discomfort of starving? Now, why on earth would anyone want that? Two cases came to public attention in the 1980s; they are not really cases in "health care ethics," but comments on a larger society:

> *Elizabeth Bouvia*: Elizabeth was born in 1958 with cerebral palsy. She was almost totally paralyzed, having the use of only the muscles in her right hand (with which she guided her wheelchair) and enough of the muscles of her face and jaw to speak and to eat. She was periodically abandoned by parents who never seemed to get their acts together. She was capable of enough organization to tap into state aid to live independently with a live-in aide; she got her high school and college degrees, even enrolled in a Master's Program. She could do some volunteer work, but never found employment. She married an ex-convict, even became pregnant, but the child miscarried, and her husband deserted her. By this time she was also suffering from degenerative arthritis, a very painful condition. In 1983, she wheeled herself into an emergency room and announced that she wanted to die ("to be free of my physical disability and mental struggle to live"). The hospital refused to cooperate with a scheme to administer her sedatives while she starved to death.
>
> The first of several court cases pitted the ACLU against the hospital lawyers—who argued not for Elizabeth's life but for their own legal immunity in their refusals to go

along with her wishes. While Elizabeth and her lawyers argued for her right to determine her own destiny, the hospital was afraid of a lawsuit (from whom?) alleging wrongful death, and the advocacy groups for the disabled, who joined the suit (and picketed outside) were afraid that her death would set a precedent for the abandonment of the disabled. For the next few years, a series of courts, advocacy groups and hospitals tried to use Elizabeth's case to set social policy in accordance with their views of what was right; meanwhile Elizabeth herself went back and forth between electing to eat and trying to starve herself. Finally some intelligent advocates managed to get her enough pain relief so her pain was no longer disabling in itself, and a good attorney managed to get her state aid channeled so that she could live independently. In 1996 she appeared on a "60 minutes" episode, declaring that she was not unhappy; in 2002 she was still alive.

Larry McAfee: Larry was a 29-year old engineering student at Georgia State University in Atlanta, when he suffered a crushing injury to his neck in a motorcycle accident and became a C2 quadriplegic (significantly more seriously paralyzed than Elizabeth). There was nothing anyone could do to heal him. His travels through the health care system seem curiously dictated entirely by costs: when he was first injured, and had what seemed plenty of medical insurance, he stayed in a specialized spinal facility in Atlanta; then moved to an apartment attended by nurses; then when the insurance ran out he went on Medicaid, and Georgia transferred him to Ohio where he could be taken care of until Georgia found a bed for him; Ohio sent him back to an emergency room in Georgia; finally a Georgia nursing home took him. He petitioned the probate court to let him acquire lethal drugs to kill himself (he'd developed a drug-delivery system that he could operate with his breath; remember, he was an engineer), and after agonizing for awhile, the court agreed to allow that. Then Larry changed his mind; shuttled in and out of group homes and other facilities, depending on what funding was available; finally ended in a group home that Georgia reluctantly funded. In 1993 a failure of his catheter caused urine to back up, which shot his blood pressure up, and caused some devastating strokes; in 1995, having lapsed into a coma, he died.

What went wrong in these cases? In both cases the paralytics surged into life—intelligent, social, effective life—anytime anyone paid attention to them, gave them a reason to live, enlisted them in any activity at all, whether or not advantageous to them. People are not meant to be abandoned. People do not thrive or flourish in useless solitude and inactivity. By now you'd think we'd know that. Larry McAfee and Elizabeth Bouvia did not need the "right" to commit suicide, they did not need (very much) medical care, and they certainly did not need hundreds of courts and lawyers and advocacy groups. What they needed was someone to partner with and something to do, just like the rest of us, and until our system figures that out, we will have to deal with these "hopeless" cases of loneliness, defeat, and desperate efforts to make the courts take the place of the community and employment that they deserve.

1.8 Physician Assisted Suicide

Jack Kevorkian: After an uneventful career as a pathologist, in 1989 Dr. Jack Kevorkian built a "suicide machine" from about $30 worth of scrounged parts, at the kitchen table of his Royal Oak, Michigan, apartment. The machine was an affair of linked IV tubes; Kevorkian would start the IV, connected to a harmless bag of saline solution, and hand the client a button-operated switch. If his client pushed the button, a valve would open, and the IV would become a powerful sedative, followed in a short time by a lethal dose of potassium chloride. Why did he make the machine? He was convinced that people had a right to die when they wanted to, and that to provide a painless and convenient way to bring about their own death was to provide a service genuinely needed. On June 4, 1990, the machine was first used to help Janet Adkins, a 54-year-old Portland, Oregon, woman with early-diagnosis Alzheimer's disease, end her life. He told the newspapers about it; the next morning a picture of that suicide machine showed up on the front page of *The New York Times*. That was too much for Michigan; Kevorkian was arrested and eventually tried for murder. Before the trial, on June 8, an Oakland County Circuit Court Judge enjoined Kevorkian from aiding in any suicides. On December 12, 1990, District Court Judge Gerald McNally dismissed the murder charge against Kevorkian in death of Adkins.

Kevorkian went on to attend other deaths, brought about either by his suicide machine or (especially after his license to practice medicine was yanked in November, 1991, and he no longer had access to lethal drugs) by another machine that rigged a facemask to the exhaust of his van, killing by carbon monoxide. Fifteen deaths later, the state of Michigan passed a ban on assisted suicide, to take effect March 30, 1993. At no time did Kevorkian conceal what he was doing; on the contrary, he defended it as a needed service not available in this country in standard medical facilities. "Right-to-Life" organizations hounded him, prosecutors kept hauling him before Grand Juries, but his technique was so simple it was very hard to stop him. In August, 1993, still in Michigan, he helped Thomas Hyde to end his life; hours after a judge ordered him to stand trial in Hyde's death, Kevorkian helped cancer patient Donald O'Keefe, 73, to end his life in Redford Township, Michigan; in October of that year, he assisted the death of Merian Frederick.

For those two deaths Kevorkian was arrested, told to post bonds of $20,000 and $50,000 respectively. He promptly refused to post bond and started to fast. He fasted until December 17, 1993, when he was released on $100 bond with a promise not to assist in any more suicides until state courts resolve the legality of his practice. On January 27, 1994, a Circuit Court Judge dismissed charges against Kevorkian in two deaths, becoming the fifth lower court judge in Michigan to rule that assisted suicide is a constitutional right. The Hyde case was reinstated; but on May 2, 1994, a Detroit jury acquitted Kevorkian of charges that he violated the state's assisted suicide ban in the death of Thomas Hyde. Eight days later, the Michigan Court of Appeals struck down the state's ban on assisted suicide on the grounds it was enacted unlawfully.

We're not done yet. On November 26, 1994, hours after Michigan's ban on assisted suicide expired, 72-year-old Margaret Garrish died of carbon monoxide poisoning in her home in Royal Oak, apparently with Kevorkian's help. She had arthritis and osteoporosis. On December 13, the Michigan Supreme Court upheld the constitutionality of the 1993-94 ban on assisted suicide and also ruled assisted suicide illegal in Michigan under common law. The ruling reinstated the cases against Kevorkian in four deaths. Nothing daunted, Kevorkian opened a "suicide clinic" on June 6, 1995, in an office in Springfield Township, Michigan. Erika Garcellano, a 60-year-old Kansas City, Missouri, woman with ALS, was the first client. The owner of the building evicted him. Ordered to stand trial for two of the suicides in 1991, on September 14, 1995, Kevorkian arrived at the Oakland County Courthouse in Pontiac, Michigan in homemade stocks with ball and chain.

Part of the tide was turning. On October 30, a group of doctors and other medical experts in Michigan announced its support of Kevorkian, saying they would draw up a set of guiding principles for the "merciful, dignified, medically-assisted termination of life." A few months before, Oregon had passed a "Death with Dignity" act permitting physicians, under certain circumstances, to prescribe lethal drugs for patients. On February 1, 1996, the *New England Journal of Medicine* published massive studies of physicians' attitudes towards doctor-assisted suicide in Oregon and Michigan. The studies demonstrated that a large number of physicians surveyed supported, in some conditions, doctor-assisted suicide. On March 6, 1996, the 9th U.S. Circuit Court of Appeals in San Francisco ruled that mentally competent, terminally ill adults had a constitutional right to aid in dying from doctors, health care workers and family members. It was the first time a federal appeals court had endorsed assisted suicide.

In 1999, however, Kevorkian crossed over a line not previously violated, by injecting a lethal drug himself into a client too weak to operate machines himself. That might not have got him in trouble, but Kevorkian had the death videotaped, and supplied a copy of the tape to the TV newsmagazine "60 Minutes," challenging the legal system to stop him. It did. For that death he was convicted of murder, sentenced to 10 to 25 years in prison, eligible for parole after eight years. No one thought the aging Kevorkian would survive jail, but in June 2007 he walked out of jail a free man. Until his death in June, 2011, at the age of 83, he continued his crusade for legalization of voluntary death.

The issue is not yet settled, to understate the matter.

Timothy Quill and Diane: In March of 1991, Dr. Timothy E. Quill published a report in the *New England Journal of Medicine* describing how he prescribed the barbiturates that a 45-year-old female patient needed to kill herself after she refused treatment for a severe form of leukemia. The arduous and painful treatment would have given her only a one-in-four chance of survival, and she rejected it; afterward, he helped her when she said she wanted to end her life. Reporters covering the story emphasized that although a few other doctors have described how they sped up the death of an incurably ill patient or helped such individuals kill themselves, Dr. Quill's account seemed to answer many of the ethical and moral objections that had been raised in previous well-publicized cases.

In particular, Dr. Quill said he had cared for the woman, whom he identified only as "Diane," for many years. He said he knew she was mentally alert and making her decision calmly. Together, doctor, patient and family had thoroughly discussed treatment and suicide options. Diane repeated her wish to die after the blood cancer began causing constant pain. He advised her to get information from the Hemlock Society, a group that advocates the right to die. "A week later she phoned me with a request for barbiturate for sleep," Dr. Quill wrote. "I knew this was an essential ingredient in a Hemlock Society suicide. I made sure that she knew how to use the barbiturates for sleep, and also that she knew the amount needed to commit suicide." Diane said tearful goodbyes to her closest friends and to Dr. Quill. Two days later her husband called Dr. Quill to say that Diane said her final goodbyes to him and their college-age son that morning and died peacefully after asking them to leave her alone for an hour.

Immediate commentary on the case was generally supportive. Dr. Ronald E. Cranford, a neurologist and medical ethicist at the Hennepin County Medical Center in Minneapolis, said, "This is a very important case, and people will have trouble criticizing the procedure." No one knows how often doctors help patients commit suicide in the United States. In an interview, Dr. Quill said he believed that many doctors had done what he had done but not talked about it and that many other doctors who might be willing to aid a suicide fear to discuss the issue with their patients. But he said it was important to tell the public and the profession that it was right for doctors to help patients they knew well to maintain their dignity in dying, including assisted suicide. He said the benefits of making the case public far outweighed what lawyers and experts told him was a small risk of prosecution or professional discipline. He had told the medical examiner that Diane died of acute leukemia but withheld information about barbiturates to spare the family any possible police investigation.

Even doing things right does not confer legal immunity. Howard R. Relin, the Monroe County District Attorney, said he would investigate the case and discuss it with the medical examiner, who was out of the country, to determine whether to submit the case to a grand jury. In New York, people convicted of aiding in a suicide can be sentenced to up to four years in prison, Mr. Relin said. But he said he knew of no prosecutions of doctor-assisted suicides in his county. "These are very difficult cases because the law is in conflict with people's perception of their right to die," Mr. Relin said. The State

Medical Society of New York, to which Dr. Quill belongs, has a policy that says, "The use of euthanasia is not in the province of the physician"; 26 states have laws prohibiting doctor-assisted suicides. But physicians and the public found it much easier to approve of Timothy Quill, primarily because Quill knew his patient well and had had a long time to talk over her options with her and her family—in contrast to Jack Kervorkian, who simply seemed to supply death over the counter on demand. In the end, Quill was not prosecuted for Diane's death.

By now, some states and nations have tentatively approved such assistance.

The Oregon experience: On October 27, 1997 Oregon enacted the Death with Dignity Act, which allows terminally-ill Oregonians to end their lives through the voluntary self-administration of lethal medications, expressly prescribed by a physician for that purpose. Under the law, two physicians must have declared the patient to be terminal, and the patient must have requested the drugs "repeatedly," three times. The law also requires the Oregon Department of Human Services to collect information about the patients and physicians who participate in the Act, and publish an annual statistical report. The reports tell an interesting story. During 2007, for instance, 85 prescriptions for lethal medications were written under the provisions of the DWDA (compared to 65 during 2006). Of these, 46 patients took the medications, 26 died of their underlying disease, and 13 were alive at the end of 2007. As in prior years, most participants were between 55 and 84 years of age (80 %) white (98 %), well educated (69 % had some college), and had terminal cancer (86 %). They tended to reside in urban areas; all patients had some form of health insurance: 65 % had private insurance, and 35 % had Medicare or Medicaid. Why did they choose assisted suicide? Their concerns were familiar: they feared loss of autonomy (100 %), decreasing ability to participate in activities that made life enjoyable (86 %), loss of dignity (86 %), and despite the fact that most were enrolled in Hospice at Home care, many were worried about inadequate pain control.

Several aspects of the Oregon law stand out in these reports. First, that's a very small number. Popular fears of widespread "euthanasia" were completely unjustified. Second, there's the surprising evidence that almost half the patients who chose to request (three times, by law) the lethal drugs did not take them. Some of them died too quickly. But most, quite possibly, did not "want to die." They wanted control over the last days of their lives, and just having the drugs in the refrigerator gave them that control. Third, they were all insured; hysterical warnings that the uninsured would take their lives because they did not have the money to pay for end-of-life care are clearly unfounded. (Another way of looking at that, is that the problems of the uninsured are not going to be solved by Assisted Suicide laws.) Fourth, Dr. Kevorkian was cited frequently in the arguments that led to the passage of the law, as evidence of the desperate lengths people will go to, when they have no legal options; DWDA provided an alternative.

The Oregon Law does not permit the physician to take any action that would actually bring about the death of the patient. Laws permitting this final step have been in place in the Netherlands for several years.

The Netherlands practice: On 10 April 2001, the Upper House of the Netherlands Parliament passed legislation decriminalizing physician assistance in bringing about death; "the termination of life on request and assistance with suicide" henceforth would not be a criminal offence if carried out by a physician and certain criteria of due care had been fulfilled. Since the Netherlands has been the most accepting of the practice of assisted suicide/euthanasia, it might be worthwhile looking into their experience.

In line with the rest of the civilized world, the Criminal Code of the Netherlands contains a variety of provisions prohibiting the intentional taking of human life (e.g. Articles 293 and 294). However, termination of life on request, and assistance with suicide, have been de facto permitted in certain defined circumstances by virtue of a non-prosecution agreement between the Netherlands Ministry of Justice and the Royal Dutch Medical Association; this agreement has been tacitly accepted since the 1980s. To comply with the requirements of the law, the physician must ensure that the request for termination of life or assistance with suicide is made by the patient and is voluntary, and establish that the patient's situation entails unbearable suffering with no prospect of improvement.

Procedural requirements include that the patient must be determined to be a citizen of the Netherlands, that two physicians must agree that the patient is conscious, competent, and terminal, that the patient must be suffering and faced with the prospect of prolonged suffering, that all alternatives to death have been discussed with the patient and have been rejected as unworkable in this situation. When all this has been determined, it is permitted to proceed, but only a physician may administer the lethal dose of drugs, and that this death must be reported to the coroner as a case of physician assisted suicide. The physician can refuse the request if he does not feel that it is right.

A review process is in place, should complaints be lodged about a particular instance of assisted suicide. It does not appear that this review process has been utilized; the procedural requirements are so thorough, and the general medical abhorrence of taking life (unless circumstances cry out for it) so reliable, that the system does not seem open to abuse. Accusations from "pro-life" groups that there are abuses, but we just don't know about them, seem to be baseless; accusations that "vulnerable groups," the handicapped, the elderly, the uninsured or poor, will turn up disproportionately in the rolls of the assisted suicides, have been refuted. (For one thing, the Netherlands provides universal health care for all citizens in all illnesses.)

Death is never easy to accept, and probably it should not be (Rage, rage against the dying of the light......). When bringing about someone's death begins to seem like an easy solution to a problem, we have probably lost moral focus. But death is inevitable, and if someone has to have control over the last days of life, most likely it should be the patient.

Chapter 2
The Cutting Edge

Medical science continues to advance, and all the new things—new procedures, new imaging machines, new drugs and theories—gives rise to anxious questions. Have we, this time, gone too far? The human being does not accept significant change easily, and medical science has always had the power to set the nerves on edge. At one time organ transplantation was the stuff of science fiction; now it is almost commonplace. Meanwhile, new knowledge can only come from research, experiments on humans like us or on animals like enough to us (or our pets) to worry us when they suffer. Research with human beings became much too urgent in the 1950s and 1960s; eagerness to get useful results quickly spawned abuses. These abuses led to the formation of the National Commission for the Protection of Human Subjects of Biomedical and Behavioral Research, mentioned above, whose reflections yielded some of the best public moral thinking this nation has allowed itself to enjoy, concluding with the articulation of the Belmont Principles, now central to ethical reflection not only in bioethics but in all fields of applied ethics.

2.1 Organ Transplantation

The God Committee (1962) and the kidney transplants: Kidney disease is serious and becoming more so: One in six Americans have some form of chronic kidney disease. 400,000 patients (according to eMedicineHealth) are on dialysis or have received kidney transplants; about 67,000 people die each year from kidney failure. Almost 40 % of adults over 60 years of age have some level of kidney failure. Causes of kidney failure, absent some exotic disease, include diabetes mellitus, high blood pressure, and obesity; all of these are on the increase, and the population is aging; no wonder kidney disease is up 16 % in the last decade.

For most of the human experience, there was little that could be done for a person with failing kidneys. Then in the 1940s, a machine that could accomplish hemodialysis, filtering toxins from the blood the way the kidneys do, was developed, but could only be used for a short time with any patient; it required careful reattachment to new veins and arteries with every use, and the patient's blood vessels quickly collapsed. In 1960 Belding Scribner, of the Swedish Hospital in Seattle, developed a "permanent indwelling shunt,"

L. Newton, *The American Experience in Bioethics*, SpringerBriefs in Ethics, DOI: 10.1007/978-3-319-00363-4_2, © The Author(s) 2013 a tube permanently attached to one artery and one vein, so that blood could flow continuously through the machine; the tube had a valve that could be closed between uses. Now hemodialysis could keep a patient alive indefinitely.

But very expensively. At the time Scribner developed his shunt, dialysis cost about $20,000 a year, and insurance companies would not pay it; the number of patients who could benefit from it was effectively infinite, and they would die if they did not get it. When Scribner moved his dialysis center to an outpatient setting, the cost went down a little, but not enough; and only 17 patients could be served. Who should receive treatment when not all can? Scribner, Swedish Hospital and the Kings' County Medical Board established a committee, the Admissions and Policy Committee, to make the decisions. (By this time insurance covered the patients' treatments.) The Committee set criteria for admission to treatment: the patient had to be from the state of Washington, had to be under 45 years of age, had to have insurance or otherwise be able to pay. Still they had many too many candidates. They drafted new criteria: family situation, employment, value to the community, ultimately character and personality. They did their work objectively and well; they were never accused of bias or corruption.

But still: how can one person "deserve" life more than another? The issue came to a head when dialysis and the Committee hit the front page of *The New York Times* in May, 1962. Scribner had taken a dialysis patient to a newspaper convention in Atlantic City to lobby for more media attention to kidney disease, and the Committee had been mentioned; the *Times* picked it up. Shana Alexander, a reporter for *Life* magazine, did a feature story on the Committee, which she called the "God committee," since it played a "godlike" role in deciding who would live and who would die at least in part on grounds of "social worth," the worth of a person to society. The article appeared in November, 1962; in spring 1963, the Seattle Times ran a front page story on several patients who had not been accepted for dialysis—"Will These People Have to Die?" Temporary financing for more machines was found from Seattle industry. In 1965, a television special on NBC featured the God committee, and raised popular sentiment for national funding for dialysis, along the lines of future programs like Medicare and Medicaid. By 1971, everyone in the country knew of the plight of rejected kidney patients, and after some more media dramatics, Congress enacted the End State Renal Disease Act (ESDRA), mandating the federal government to pay for dialysis for anyone who needed it. That ended the need for the God Committee.

ESDRA solved some problems, but led to others. By 2006, despite the efforts of several cost-containment mechanisms, dialysis was costing Medicare $16 billion a year. In a country with 45 million citizens unable to obtain medical insurance for even basic medical care, that seems disproportionate. And of course, sufferers from other diseases wanted to know why their disease could not also have mandatory federal funding; no good reasons have been offered. The use of media and of public awareness on behalf of health care decisions also pioneered in this case. Scribner's trip to Atlantic City was the first time the medical profession had broken out of confidentiality to enlist the public for help; inevitably, it launched the first public discussion of the terms and conditions of health care, the ethics of medical choices, and the insertion of politics into health care decisions. This last development has been a very mixed blessing (see *Terri Schiavo*, above).

Dialysis is not an ideal solution for sufferers from kidney disease; it is time-consuming, exhausting, occasionally dangerous (it is very hard to exclude completely disease-causing organisms), and a drain on the insurance. Far better would be a kidney transplant. But then, of the available recipients, who should get the kidney transplant?

The first recorded kidney transplant happened in 1945 in Peter Bent Brigham Hospital in Boston, when a kidney from a recently deceased patient was transplanted into a woman in kidney failure following the birth of her baby. It functioned badly, did little good but apparently little harm, and the woman's own kidneys recovered. In 1952, another transplant, from deceased mother to her son, functioned for 22 days until it was rejected. The first successful kidney transplant was accomplished in 1954, also at Peter Bent Brigham, between identical twins, Ronald and Richard Herrick.

Why did that one work so well? As most of us now know, the reason why we can't treat our organs as interchangeable parts, freely exchanged according to need, is that each of has a built in physiological hostility to anything recognized as strange or alien, a hostility that forms the essential center of our immune system. Without it, we're dead at an early age, invaded by any and all passing organisms. With it, we reject transplanted organs, unless we figure out some way to turn it off. Steroids will turn it off, or down, temporarily; but steroids have long-term side effects that we don't want. Various immunosuppressive drugs were used in the early days of transplants, 1950s and 1960s, but none worked very well until the discovery of cyclosporin(e) (spellings differ) in 1971. Since then the success rate of organ transplantation has soared. Identical twins, of course, have identical immune systems, and do not recognize each other's organs as alien. That's why the Herrick transplant worked in 1954.

Kidney transplants have one other advantage over other organ transplants: the kidneys are paired organs where each member of the pair can perform the whole task of filtering the blood by itself—as a matter of fact, if you remove one kidney, the other kidney will promptly expand to fill its space, and the health of the donor will not be affected. So "live-donor" organs are preferred for kidney transplants, the closer the genetic relationship to the recipient, the better. With the donor alive, the question of deterioration of the organ does not arise, as it does when the donor is deceased. With advances in immunosuppressive drugs, "cadaver-donor" organs can achieve a decent success rate, but the drugs have a universal disadvantage— they suppress the immune system, increasing the probability of lethal infection and cancer.

Live-donor sources are possible for kidneys, and the practice is extending; at the turn of the millennium, it had become standard practice to look for live donors among the relatives of those suffering from a variety of liver failures (one lobe of a healthy liver can often take the place of a failed liver, especially for children), and lung failures, like cystic fibrosis. The lungs are not exactly paired organs, as not being separated in nature, but a lobe of a lung can often be attached in a sick person's body, and take over the role of the original pair of lungs. The heart remains the most important organ that can only be obtained from cadavers.

The first heart transplant, mentioned briefly above, took place in December of 1967, before immunosuppressive drugs were fully developed. Christiaan Barnard found himself with a long-term patient, Louis Washkansky, who was clearly dying from heart failure. Talk of heart transplants was widespread in the profession, and many of the necessary techniques were known; Norman Shumway, cardiac surgeon at Stanford University, had announced a month before that he was ready to do a heart transplant when a suitable donor could be found. Then the body of Denise Darvall was brought into Barnard's hospital from an automobile accident; with her father's consent, he removed her heart, placed it in Washkansky's chest, and spent the next 17 days trying to keep him alive, an attempt that ultimately failed. But there seemed to be hope that later transplants might work; after the first week's struggles with multi-organ failure, steroids had bought Washkansky five good days, during which the heart beat well and there was real hope of recovery. (Rejection set in from that point.) Barnard tried again, with Philip Blaiberg, who may have been in better health than Washkansky at the time of the surgery. Blaiberg walked out of the hospital on his own, living nineteen months before dying from complications of heart disease. The world was elated at the possibilities, and surgeons in many nations tried transplanting hearts, livers, lungs. Most patients died before long, the transplanted organ rejected; long-term success awaited the development of much better drugs.

(In an attempt to get around the scarcity of cadaver hearts for heart patients, physicians and engineers collaborated through the 1970s to develop a "totally implantable artificial heart," a small but efficient pump that would simply take the place of the heart of flesh. Never mind the Biblical echoes, please. By 1982 Willem Kolff and Robert Jarvik of the University of Utah thought they had a model ready to go, and William DeVries, a cardiac surgeon at the medical center, implanted it in the chest of cardiac patient Barney Clark on December 1. It didn't work very well; Clark died in quite a bit of discomfort. Later recipients fared worse. One of them, William Schroeder, lived 21 months, most of them miserable, and talked to President Ronald Reagan on the telephone. He died of a series of strokes. What the physicians had not known, or had forgotten, is that the heart is much more than a pump; it is a highly sensitive chemical control mechanism, a source of hormones, responding in multiple unknown ways to stimuli from the brain, and communicating complex system adjustments to the rest of the body. Engineers will design a computer capable of taking the place of the brains of experimental surgeons before they find a machine capable of replacing the heart. Latest news is that the Left Ventricle Assist Device (LVAD), supposed to be a partial temporary artificial heart, isn't very good at keeping people alive, either; research continues and improvements are expected.)

For a recipient of a heart to live, the donor has to die. Not just any cadaver will do for an organ transplant. First, consent must be obtained. If the deceased has indicated, in a living will or other document, a willingness to be an organ donor, that solves most problems (but not all; the family can still protest). If there is no document, and only 20 % of adults have signed such, the family can be asked to allow harvest of viable organs. The usual circumstances surrounding the death of a potential donor (sudden, unexpected, violent death of a young and healthy person) make this request very difficult. Second, the organs must be healthy and strong enough to be useful to the recipient. That rules out the organs of most victims of disease, including all victims of cancer, for the disease or the cancer may well have spread to the organs. It also rules out the most willing of organ donors, the very elderly; no one wants an 86-year old heart. In practice, the eligible donors turn out to be young (under-40) victims of traffic accidents or gunshot wounds to the head. These are difficult to come by; gunshot victims do not end up in hospitals in time, and the victims of traffic accidents have steadily decreased in number, thanks to seat belts, air bags, tougher laws against drunk driving, and required helmets for motorcyclists. (Citing respect for liberty and self-determination, states occasionally repeal the helmet laws, with the verifiable effect of making many more organs available for transplant. The ethical trade-off here has not been adequately explored.) The problem created by these restrictions is that the next of kin, usually parents, have to be asked to allow organ donation as they stand by the body of their dead son or daughter, a young future brutally cut short, in the throes of shock and grief. It is a scene that no one enjoys, and many cannot bear.

How can we make more organs available for transplant? One attempt, in use in some states, requires any hospital with a possible organ donor—the brain-dead body of a young accident victim—to notify a state organ procurement team, made up of carefully trained professionals, who will arrive and conduct the interview with the next of kin. Life support for the patient could not be removed until the team arrived. The rationale for such teams is that the staff at the hospital may be unwilling to ask for organs due to the sensitivity of the situation, and the organs will be lost as the potential donor dies. The availability of a trained team may increase the quantity of organs donated. Another possibility, in practice in France and other European nations, is "presumed consent": when a viable donor comes along, whether or not next of kin are available to consult, it is presumed that the deceased wanted to donate organs, which are harvested immediately. That practice does indeed raise the number of organs available; it is not clear that such a practice would work in the United States, where trust in the medical profession is much lower.

Once the organs have been obtained, how shall they be distributed? The usual method of "distributing" medical resources is no method at all: physicians simply assign the medical resources under their control to the patients in their care, and if two patients need the same resource, the first to need it is the first served. (And the patients who have no insurance, or no access to medical care, are not served at all.) Somehow that chaotic method did not seem adequate to the problems of organ distribution: for an organ transplant to work, the donated organ must be placed in the recipient's body within a very short time after removal, well oxygenated, under near ideal operating conditions; and the recipient must have been carefully screened for medical fitness, that is ability to tolerate and support the donation.

Further, donor and recipient must be histocompatible. "Histocompatibility," as described on the website of the United Network for Organ Sharing (UNOS), the mechanism created to solve the distribution problem, is the condition in which a donor and recipient share antigens so that a graft (donated tissue) is accepted and remains functional. Histocompatibility antigens, called HLA (for *human leukocyte antigens*), are proteins on the surface of the cells in the body. Their main function is to help the immune system defend against invaders such as bacteria, viruses, and parasites. Unfortunately, the immune system can also recognize as foreign

the histocompatibility antigens of other people's cells and will fight them, causing rejection of grafts and donated organs. Because HLA antigens can be recognized as foreign by another person's immune system, transplant professionals try
to match as many of the HLA antigens as possible, between the donated organ
and the recipient. That way, there is less of a chance that the recipient's body
will reject the organ. In order to do this, the HLA type of every potential organ
recipient is determined before they are placed on the waiting list. When a potential organ donor becomes available, the donor's HLA type is determined as well.
A match program is run through UNOS and the best possible recipient for each
organ is chosen. Further tests, known as crossmatches, are performed to make
absolutely sure that the donor organ is suitable for the recipient. Above all, for the
transplant to work, someone in the system has to know where the possible donors
and recipients are—they could be anywhere, and are very unlikely to be in the
practice of a single physician. The possibilities for failure are legion, and the transplant is of infinitely high value—it means life for the recipient. We cannot leave
this practice to chance.

So a national system was necessary just to sort out the logistics of organ transplants. That system already centralizes the decision-making, and makes it visible,
but there still were no criteria to decide who gets the very scarce and very valu-
able organ. The God Committee was one answer: appoint a group of prominent
citizens, trusted by all, to decide who is most deserving of the pool of medically
appropriate recipients. But the God Committee, and the "social worth" criteria
they had incorporated into their thinking, had come under vigorous attack from
the bioethicists: Sanders and Dukeminier argued in a 1968 *UCLA Law Review*
article that social worth criteria should never have been used, saying that the *Life*
article "paints a disturbing picture of the bourgeoisie sparing the bourgeoisie, of
the Seattle committee measuring persons in accordance with its own middle class
suburban value system: Scouts, Sunday School, Red Cross." Nonconformists
seem to have no chance for treatment, as they see it; the article concludes with
the famous tag, "the Pacific Northwest is no place for a Henry David Thoreau
with bad kidneys." Following on that critique, medical sociologists Renee Fox and
Judith Swazey pointed out in 1974 (*The Courage to Fail*) that the selection criteria mirrored "the middle-class American value system," where the preferred candi-
date was characterized by "decency and responsibility," with a strong family life,
who had "demonstrated achievement through hard work and success at [his or her]
job, who went to church, joined groups, and was actively involved in community
affairs." The fact that these accounts were judged to be warranted, and devastating,
criticisms of the Committee throws interesting light on the ethical tendencies of
bioethics in the late 1960s and 1970s. (What's wrong with decency, responsibility,
hard work, church attendance, and active citizenship? If we have to select on some
criteria, we could do a lot worse than these.)

Justified or not, these criticisms ensured that no more God Committees would
show up on the American scene. Some bioethicists had suggested random distribution: that after recipients had been matched to the available organs medically and
genetically, the system should simply draw straws to see who should get the organ.

That didn't seem like a good idea, either. Congress had had a few efforts at governing organ transplantation. In 1987 these attempts, the National Transplantation Act and the Task Force on Organ Transplantation, were combined to create UNOS, which at least did away with regional competition (and hoarding of organs to reserve them for local patients), and adopted some generally fair criteria for allotting them. It created waiting lists: each medical center that was prepared to transplant organs sent a list of available recipients. When a donor became available, the organs were distributed to the recipients according to how long they had been on the list (first listed first served) and by seriousness of illness (the sickest get to the front of the line). That's probably as fair as it's going to get. There are a multitude of difficulties: a person too poor to have insurance or access to medical care will not get on a list at all; a person wealthy enough to devote full time to acquiring a needed organ can go to every center in the country and get on their list, multiplying his chances; there are centers abroad that can supply organs to the very wealthy; an influential surgeon can often put one of his patients in front of others in the line; and there is the unpredictable legacy of the "publicity" factor.

Recall, the whole problem of allotting scarce medical resources first came to public attention when Scribner brought a patient to a newspaper convention. Since then, UNOS or not, every time a patient, or his family or his physician, discovers that he needs a transplant, the newspaper is the first to know, then the talk shows, then local fundraisers to cover medical expenses. The hope is not to persuade the managers of UNOS to put the endearing child or brave mother at the front of the lines they control. The hope is that the next-of-kin of a newly deceased potential donor, seeing the appeals on the local news, will insist that the required organ be donated only to that recipient. Since the implication is that if the request is refused, the organ will not be donated, it's generally granted, restricted only, as above, by histocompatibility.

No one has figured out a way to get around the inherent unfairness of the publicity factor—nor the similar distortion of the public figure. Consider the liver transplant: it is the most difficult of the transplant surgeries, requiring highly skilled teams to make it work at all. Considerations of allotment of livers, among the scarce organs, are complicated by the fact that the most common reason why livers fail is alcohol abuse. Alcohol-related end-stage liver disease (its own diagnosis: ARESLD) is the condition most likely to raise the need for a transplant, but it raises a serious question of justice alongside: this patient did this to himself. Does he deserve an organ transplant after destroying his own liver? While this debate continued, beloved retired Yankees' baseball star Mickey Mantle showed up at his hometown hospital with ARESLD and liver cancer. He was put on the UNOS list, and it seemed a mere matter of days before he got a donated liver. Was that fair? UNOS insisted its guidelines had been followed, but its guidelines insist that cancer is a disqualifying factor; in fact, after receiving the transplant, Mantle died in a matter of weeks from cancer.

When we take up the larger problems of justice in health care, we will have to confront the general problem of the responsibility of the individual to care for his own health and the appropriate response of the health care system to evidence that he has not. Should smokers be charged more for surgery—or refused treatment altogether? At least, should smokers and heavy drinkers be declared ineligible for heart transplants? This question may not be resolved in our generation.

2.2 Problems in Research

Much medical research is done on animals. Do we have any right to cage and abuse animals in this way? Does it matter if the animal suffers extensively, or just participates in the experiment and is then killed? One of the first targets of the reformers was the Draize test:

> **The Draize Test**: Humans like to rub, or spray, cosmetics on their faces and hair, and some of the cosmetic substance can get into the eyes. For years, the manufacturers tested their new cosmetics by introducing them into the eyes of rabbits, after clipping the rabbits eyes open to prevent self-defense and clamping their heads into stalls so that they could not move. If the rabbits' eyes did not seem to be badly affected, it was concluded that the substance was safe enough for human use around the eyes. The test was terribly painful for the rabbit, and eventually blinded them. Was this necessary? Much fruitful work is being done to develop cloned tissue models to replace the rabbits.

Then the reformers pointed out that beyond the individual procedures, the whole method of counting success was unnecessarily cruel:

> **LD-50**: How toxic is a substance—a new agricultural product, a new floor cleaner, especially new pharmaceuticals? A standard way of discovering the toxicity of something, is to feed it to (or spray it on) a large population of test animals, mice or rats or rabbits or beagles. The question asked in this test is, what quantity of the substance is sufficiently toxic to bring about the deaths of 50 % of the test animals? That's quite a bit of animal death, and the ones that do not die, of course, are very sick and miserable. Is there any other standard that can be used? By reformulating their counts, it seems to be possible to use an LD-10 standard instead, which will not only kill off fewer of the test animals, but will leave the remainder in much better shape.

As experimentation with animals began to accumulate publicity, a new generation of activists closed in on the research to gather the materials for public information:

> **Edward Taub's "somato-sensory deafferentation" studies in monkeys**: Alex Pacheco, an activist for PETA (People for the Ethical Treatment of Animals) volunteered in Edward Taub's neurology laboratory in 1981, pretending to be a neurology student, but really in order to videotape the activities of the lab. Taub had been cutting nerves to the limbs of his subject monkeys to simulate the injuries suffered in brain injury and stroke. He hypothesized that one reason that people had such trouble regaining the use of their limbs after a stroke or brain injury was that since the affected limb did not respond in any way they were used to, they essentially gave up on it, and moved all activity to the uninjured limb, a situation he called "learned helplessness." He thought that if the good limb were restrained—tied down—during the rehabilitation period, the brain might rewire itself to recover the use of the affected limb. The experiments cannot have been pleasant to watch, and Pacheco got some damning film to show the world. To the whir of TV cameras, police raided Taub's laboratory, while PETA handed out literature. Taub was briefly convicted of failing to give the animals proper veterinary care; the conviction was reversed; he went on

to become one of the most admired neurologists in the field, as it turned out that he had indeed brought about more recovery of function in the monkeys than any had thought possible. His Constraint-Induced Movement Therapy (CI) is now widely used, not only for stroke victims, but for military personnel returning from Iraq with traumatic brain injuries, and is widely praised for obtaining marked increase in their ability to use paralyzed limbs.

2.3 Philosophical Issues of Animal Research

What right do we have to do research with animals anyway? How are animals like ourselves? How different? What is important—their reason, their communication, or their ability to suffer? When we do research with humans, we have to get consent from the subject; how do we get consent from an animal?

There are two major lines of thought to protect the animals: Utilitarians concentrate on the balance of pleasure of pain, claiming that the central issue is the suffering of the animals. This line is argued by Jeremy Bentham and Peter Singer, who are focused on what should be done for the animals' welfare. The other line, led by Tom Regan and others, holds that all creatures who are subjects of a life—a life that can go better or worse, whether or not they are aware of it—have a right to be left alone that that life may flourish; this is an animal rights argument.

Is there some obligation to let animals live in their own place, their own natural habitat, according to their own laws? If there is such a general duty, does it follow that there is an obligation to release to the wild all restrained or tamed animals? Would that be good for them? If not, are we obligated to keep the animals we have? Or may we humanely dispose of them? Meanwhile, a new moral issue is raised by the tactics of Animal Rights activists. Is there a right to destroy laboratories, burn data, when animals are being made to suffer? Who authorized this? Whatever our sympathies, must we not hold animal rights activists responsible for the harm that they do?

As a practical matter, if we do not decide to abolish research with animals altogether, can we adopt guidelines to minimize the harm done? If we decide that the enormous value of using animals for research—there is very little we know about medicine, surgery, or pharmaceuticals, which we did not discover through tests or practice on animals—outweighs the claims animals have to be left alone, we will still have the obligation to minimize the suffering of the animals in our labs. The morally responsible program seems to involve three imperatives:

1. **Replace**: to the extent possible, replace the use of live animals with lab-cultured human tissue, grown from surgical detritus available from hospitals.
2. **Reduce**: to the extent possible, reduce the total number of animals used in the tests.
3. **Refine**: modify investigative procedures to cause less pain and suffering, for a shorter time, leaving the animals healthier; or euthanize them earlier.

2.4 Research with Human Subjects

If we must conduct research with human subjects, how can we do it ethically? Do the subjects have rights that should be protected? The question had been a staple of science fiction for a century and more, but did not arise in any serious way until the end of World War II, when the documents from the concentration camps showed us that many Nazi physicians had used concentration camp populations for experiments that might have been of use to the German military—on, for instance, the best way to revive flyers who had ditched in the waters of the North Atlantic, or the human tolerance for very low atmospheric pressure. Most of the unwilling subjects died, or were severely injured. The cruelty of the experiments shocked the world, comprising a good part of the case of "crimes against humanity" in the Nuremberg Trials that followed the war. The international community adopted a set of "Nuremberg Principles" to govern human subjects research, requiring the voluntary consent of each subject, complete information of risks given to the subjects, overall, a careful balancing of risks and benefits (not so easy, since the subjects took the risks and the patient population as a whole reaped the benefits), and the right to withdraw from the experiment at any time. It further specified that the question the research was exploring had to be worthwhile, and that the investigators had to be qualified to carry on the research and understand the results. The Nuremberg Code became the ethics guide for human subjects research everywhere—not always observed.

One of the first cases of human subjects research in the United States that came to our attention was the Tuskegee Syphilis study. It's important enough to include in detail.

The Tuskegee Syphilis Study: In the late 1920s, the U. S. Public Health Service undertook a study of the occurrence of the venereal disease syphilis among the male African American employees of the Delta Pine and Land Company of Mississippi, and found an appalling prevalence of 25% of 2000 men tested. The PHS brought in the Julius Rosenwald Fund, a health-oriented charity based in Chicago, to treat these employees, and later expanded the treatment protocol to include five more counties in the South. In 1928 the PHS and Rosenwald decided to collaborate on a study of the course of syphilis in men and the effectiveness of the existing treatments.

The "treatment" for syphilis from 1900 to 1940 was an amalgam of heavy metals—mercury, arsenic and lead—called "salvarsan." When it became clear that salvarsan was killing people, it was replaced by "neosalvarsan," a synthetic drug known as an "organoarsenic" compound. It became available in 1912, was equally good at suppressing the symptoms of syphilis, and superseded the more toxic and less water-soluble salvarsan as an effective treatment for syphilis. It too had bad side-effects; incidentally, the side-effects of mercury poisoning are the same as the medical description of the last stages of syphilis. Even as the protocol started, no one really knew whether the drugs being given for syphilis were doing more harm than good.

Meanwhile, some intriguing work by Norwegians Boeck and Bruusgaard between 1890 and 1930 suggested that maybe syphilis should be untreated altogether—that those who had had syphilis for 30 or 40 years seemed to be symptom free, and of those who were clearly positive for syphilis, many never developed the symptoms at all. So as the protocol started, no one really knew if it was necessary to treat syphilis at all.

The setup of the study reflects the strong suspicion that syphilis followed a different course in blacks than in whites. The speculation was undoubtedly tinged with racism, but not implausible: people who look radically different from ourselves often share a different genetic heritage and show up with different diseases or diseases in different forms: from Tay-Sachs disease, to sickle cell anemia, to the alarmingly heightened prevalence of hypertension in the African American population. On the other hand, most diseases are not different among us. So as the protocol started, no one really knew whether to look for different effects, black and white.

The original protocol called for six groups, matched by uncertainty:

a. Black, syphilitic, treated	b. Black, syphilitic, untreated	c. Black, not syphilitic
d. White, syphilitic, treated	e. White syphilitic, untreated	f. White, not syphilitic

Comparing a/d with b/e, they might find out if treatment did any good over the long run, or maybe did more harm than good;

Comparing b/e with c/f, they could find out if syphilis untreated really shortened the lifespan, or if treatment might safely be foregone;

Comparing a/b/c with d/e/f, they could find out if there was a real difference between blacks and whites.

> Let us continue with the study. Early in the study period, the Great Depression hit the research community. The Rosenwald Fund had to withdraw its support. Without the Rosenwald Fund, the PHS did not have the resources to implement treatment (they could not purchase the drugs). White subjects, with or without syphilis, could not be recruited— there was no money to compensate them. Dr. Taliaferro Clark of the PHS suggested that the project could be continued with group (b) alone, the documented observation of African American men with syphilis, not given any treatment at all. At least they could find out if syphilis shortened the life by comparing the experimental group with the surrounding population. So that's what they did.
>
> With Rosenwald out, PHS enlisted the support of the Tuskegee Institute, a prestigious and trusted college for African American students, whose reputation would recommend the study to the African American community of the area. In return, Tuskegee Institute received money, training for its interns, and employment for its nurses; it was joined by black church leaders and other community leaders, including plantation owners, who also encouraged participation.

For the participants, the project seemed to be a reasonably good deal. At this time, African Americans had very little access to medical care. For many of the subjects, the examination provided by the PHS physician was the first medical examination they had ever received. Along with free examinations, food and transportation were supplied to participants; burial stipends were used to get permission from family members to perform autopsies on study participants. The controversy over "Tuskegee" comes from the information supplied, or rather denied, to the participants.

> While study participants received medical examinations, none were told that they were infected with syphilis; the reason for the examinations and the "treatments" they were

receiving was held to be "bad blood." PHS officials not only denied study participants any treatment there was, but also prevented other agencies from supplying treatment. As World War II gathered steam, many participants were drafted, and after blood tests, ordered to be treated for syphilis; the PHS managed to block the treatment. In 1943, the PHS began to administer penicillin to patients with syphilis, but study participants were deliberately excluded. Through cooperating local health departments, the PHS managed to keep penicillin from the subjects until 1972, when a whistleblower, PHS physician Peter Buxton, broke the story to the Associated Press. Incredibly, the study continued even after its details were splashed all over the nation's newspapers.

The popular reaction to the Tuskegee Study—astonishing to the investigators who had carried it out—was outrage that heartless scientists could take this population of vulnerable people and turn them into lab rats (or guinea pigs) just to advance Science (and their own careers). Comparisons to the horrendous "experiments" of the Nazi concentration camps surfaced immediately. The public demanded action from their lawmakers, and got it. The Congressional subcommittee on Health, Education and Welfare, chaired by Senator Edward Kennedy, held hearings on clinical research in 1973, resulting in a rewrite of the Health, Education, and Welfare regulations on working with human subjects. Turning to the private law, the study subjects filed a $1.8 billion class action suit in U.S. District Court, also in 1973; in December of 1974, the U.S. government paid $10 million in an out of court settlement. Nor did the outrage subside at that time: years later, when the plague of HIV-AIDS hit the African American community, public health workers had a great deal of trouble getting African American men to consult physicians, because of suspicions (and superstitions) traceable to the Tuskegee scandal. Tuskegee still stands as a paradigm case of an immoral experiment. Is this entirely just?

Let us consider. When Rosenwald pulled out, and the federal government pulled most of its money out of the project, a few stalwarts were left to decide what to do. They decided that since they had no more money for treatment, groups a & d were out; since they had blessed little money to follow up on people who were off to California as fast as they could go, following groups c & f would not be practical; and all of the people in group e were off to less drought and dust-stricken parts also. But it would be possible to follow group b, mostly sharecroppers who were stuck in the area; since many of the members of the crew had already contributed time and effort, it was decided to keep the project going just following group b. Was that research unreasonable? Not necessarily *at that time*. Given the outcome—that members of group b were still alive to receive compensation from President Clinton, while no one originally from groups a and d came forward—it would seem that they demonstrated in fact that the treatments available at that time, amalgams of heavy metals, probably did do more harm than good. That justification disappears in 1943, when penicillin became available—no matter how much they wanted to find out if syphilis needed to be treated at all, when they had a safe and effective remedy, they should have used it, and the PHS's efforts to keep their subjects away from penicillin was inexcusable.

Was anything learned in the Tuskegee Study? Yes, that the medications that incorporated heavy metals did more harm than good. But more importantly, the reaction to the study alerted the scientific community to the importance of two factors that they had completely ignored—that subjects had a right to be informed about the nature and purpose of the study, and of the risks to which they would be subjected, and that issues of social justice—of the vulnerability of their subjects, especially vulnerability resulting from differences of race, gender, and socioeconomic status, should be considered in the design of the study.

There were other cases of human subjects research that caused controversy. In 1963, a cancer team at the Jewish Chronic Diseases Hospital injected cancer cells into a goodly number of elderly demented patients. The point was to see if the elderly body would reject the cells in the same way younger bodies had been shown to do. The investigators claimed that all the patients had been asked if they would like to participate, and all had agreed. Of course they had omitted to mention that the injection was of cancer cells, for they did not want to alarm the patients. Those who reviewed the study felt, correctly, that omitting the information about the content of the cells made "informed consent" a sham. The investigators were disciplined, and in the end, no patient contracted cancer. Then, in 1966, Henry K. Beecher of Harvard Medical School documented 22 cases (down from 32 in an earlier version) of apparent abuse of subjects published in the medical journals of the time; his article, published in *The New England Journal of Medicine,* brought the medical profession to its feet and evoked a call for much more careful regulation of clinical research. Ironically, neither the Tuskegee study, the Jewish Chronic Diseases Study, nor the Beecher article actually precipitated federal action to regulate research. That stimulus came from opposition to fetal research, which had begun with the *Roe v. Wade* decision, establishing the permissibility of abortion (and therefore of the availability of fetuses to study). In the wake of that decision, several lines of research immediately suggested themselves, especially the testing of drugs to discover whether they crossed the placenta in pregnancy. From 1957 through 1960, a drug called thalidomide was sold as a sedative effective in pregnancy. Unfortunately, it turned out to be a teratogen; if taken when the fetal limbs were forming, the child was likely to be born with grotesquely deformed limbs. No one wanted a repeat of that. So after *Roe v. Wade,* a woman seeking an abortion could earn a small stipend by taking an experimental drug before the abortion and consenting to the examination of the fetus afterward to find out if it contained traces of the drug. If it did not (in repeated trials), then the drug did not cross the placenta and was safe for pregnant women to take. When the political furor over abortion extended to fetal research, the U.S. Congress was pressured to take up the cause of regulation of research with human subjects, which it did with the 1974 formation of the National Commission for the Protection of Human Subjects of Biomedical and Behavioral Research, a committee specifically charged to come up with sensible regulations in an ethically problematic field, research with human subjects.

The Commission did its work well, and in April of 1979 issued the Belmont Report, summarizing the principles that seemed to have been at work in the decisions on regulation: beneficence (including non-maleficence), justice and respect for persons. These have been the foundation, not only of the ethics of human subjects research, but of most of the fields of applied ethics, ever since.

What happens when the part of the human subject that is being investigated is his mind, his opinions, his set of beliefs that prepare him to work in the world? Consider the famous Milgram investigations into obedience to authority.

> ***Stanley Milgram and the Research on Obedience to Authority***: When the volunteers arrive to participate in Milgram's study, he puts them in teams of three: a "teacher", a "learner," and of course the "experimenter," in charge of the study, a biologist in a white coat, and an electric generator with wires. Only the "teacher" is an actual participant, unaware of the arrangements for the experiment; the "learner" is a confederate of the experimenter, an amateur actor trained to act for the role. The experiment is described to teacher and learner as part of a study of memory and learning: the study is designed to find out if "negative reinforcement," i.e. punishment, is helpful in getting people to learn. The participant and the actor are then told to draw slips of paper to determine who will take the role of teacher and who the learner; in actuality, both slips of paper are marked "teacher." (The actor claims that his says "learner.") At this point, the "teacher" and "learner" were separated into different rooms where they could talk to each other, but not see each other. (In one version of the experiment—it had multiple iterations—the actor mentioned to the participant that he had a heart condition.)
>
> The teacher, the actual subject of the experiment, was given a 45-volt electric shock from the generator as a sample of the shock that the learner, wired to the generator, would receive during the experiment. The teacher was then given a list of word pairs to teach the learner. After reading the whole list to the learner, now hidden behind a screen and apparently wired to the generator, the teacher would then read the first word of each pair and read four possible answers. The learner would press a button to indicate his response. If the answer was correct, the teacher would read the next word pair. But if the answer was incorrect, the teacher would administer a shock to the learner, with the voltage increasing for each wrong answer. (In reality, there were no shocks, of course. After the actor disappeared behind the screen, he set up a tape recorder integrated with the electro-shock generator, which played pre-recorded sounds—grunts, groans, increasing to screams of pain—for each shock level. The use of the pre-recorded sounds made sure that each subject heard exactly the same reactions, to make the results comparable.) After a number of voltage level increases, the actor started to bang on the wall that separated him from the subject, and, in some iterations, complain about his heart condition. After several shocks of increasing ferocity, all responses by the learner would cease.
>
> Many of the subjects, ordinary citizens all, worried about the pain apparently felt by the learner, and questioned the purpose of the experiment and asked permission to stop. Most continued after being assured that they would not be held responsible. Some exhibited signs of extreme stress as the screams of pain increased. But if the subject expressed a desire to halt the experiment, he was given a succession of verbal prods by the experimenter, in this order:
>
> 1. Please continue.
> 2. The experiment requires that you continue.
> 3. It is absolutely essential that you continue.
> 4. You have no other choice, you *must* go on.
>
> If the subject insisted on stopping at this point, the experiment was halted. If he continued, it was stopped after the subject had given the maximum 450-volt shock three times in succession. Had these shocks been real, the "learners" would have been dead.

Would we, any of us, have administered shocks we thought were lethal to a total stranger, just because the professor in a white coat told us to? Before conducting the experiment, Milgram polled Yale University psychology majors as to what they thought the results would be, and went on to poll his colleagues in the psychology department. All of the poll respondents believed that only a few subjects would be prepared to inflict the maximum voltage. But in Milgram's first set of experiments, 65 % (26 of 40) of experiment participants administered the experiment's final 450-V shock, though many were very uncomfortable doing so; at some point, every participant paused and questioned the experiment. Some said they would refund the money they were paid for participating in the experiment! But no participant steadfastly refused to administer shocks *before* the 300-V level. Later, Milgram and other psychologists performed variations of the experiment throughout the world, with similar results, although unlike the Yale experiment, resistance to the experimenter was reported anecdotally elsewhere. The only consistent variation in results stemmed from the place the experiment was conducted: the greater the prestige and respectability of the locale (Yale laboratory vs. office in the inner city), the higher the level of obedience.

Later analyses of the results of the experiment show a remarkably consistent percentage of subjects willing to inflict the final voltages, 61–66 %, regardless of time or place. (There is a little-known footnote to the Milgram Experiment, reported by Philip Zimbardo: none of the participants who refused to administer the final shocks insisted that the experiment itself be terminated, nor left the room to check the health of the victim without requesting permission to leave, as per Milgram's notes and recollections, when Zimbardo asked him about that point.)

The Milgram Experiments were promptly criticized on ethical grounds, because of the extreme emotional stress suffered by the participants, during and after their participation. Had the subjects been asked beforehand if in obedience to orders they would hurt or kill a stranger, they would surely have said they would not. But look, when the choice was theirs, they did it. They are not the persons they thought they were; the kind of suffering they went through has since been called "inflicted insight." Yet despite the suffering, 84 % of former participants surveyed later said they were "glad" or "very glad" to have participated.

The experiments provoked emotional criticism more about the experiment's implications than with experimental ethics. Joseph Dimow, a participant in the 1961 experiment at Yale University, wrote about his early withdrawal as a "teacher," suspicious "that the whole experiment was designed to see if ordinary Americans would obey immoral orders, as many Germans had done during the Nazi period." In fact, that was one of the explicitly-stated goals of the experiments. In the preface to *Obedience to Authority,* the book recounting the experiments, Milgram had written "The question arises as to whether there is any connection between what we have studied in the laboratory and the forms of obedience we so deplored in the Nazi epoch."

It is unfashionable to defend the Milgram experiments at this time. But it should be pointed out that just prior to these studies, in the 1950s, the psychological literature had been full of speculation on "the authoritarian personality," a personality type that was supposed to predispose its owner to obey authority unthinkingly, jump off a cliff if commanded, that sort of thing. The nation was in agreement that Germans had such personalities, which is why they could imprison and abuse Jews and carry out unlawful executions. Americans, we were told, had "democratic personalities" and would *never* obey unjust or unreasonable or harmful orders. This literature, part of a decades-long effort to explain the Third Reich to the stunned nation, was ended by the Milgram experiments; Americans, it turned out, were very likely to engage in hurting people on command, just like Germans. And the Americans in question were not soldiers, trained to obey, for whom disobedience would mean instant death, but ordinary citizens under no compulsion stronger than the instructions of an unarmed man in a white coat.

> **The Stanford prison experiment,** conducted in 1971 by social psychologist Philip Zimbardo at Stanford University, was supposed to find out how prisoners and prison guards adapted to their roles in any prison. Twenty-four male undergraduates were selected, carefully screened for mental stability and general health, and assigned the roles of guards and prisoners in what was described as a "two-week prison simulation." They were to be paid $15 per day for the two weeks. The study took place in a mock prison in the Stanford psychology building. Zimbardo himself took on the role of "superintendant," while his research assistant became the "warden." Contrary to the predictions of their professors, the participants rapidly took on their roles. The guards became authoritarian to the point of sadism, while the prisoners exhibited all of the stress symptoms that are found among all victims in institutional settings. After sensing that everyone had been too absorbed in their roles, including himself, Zimbardo terminated the experiment after six days.
>
> The experiment had set out to test the idea that the inherent personality traits of prisoners and guards led to observed abusive prison situations. In order to duplicate what he took to be the "dehumanizing" characteristics of prisons and other total environments, Zimbardo provided the guards with weapons, uniforms, and mirrored sunglasses to prevent eye contact.
>
> Prisoners wore ill-fitting smocks and stocking caps. Guards called prisoners by their assigned numbers, sewn on their uniforms, instead of by name. A chain around their ankles reminded them of their roles as prisoners. They were fingerprinted, photographed, then taken to their mock prison where they were strip-searched and given their new identities.
>
> The experiment quickly got out of hand. Prisoners suffered—and accepted—sadistic and humiliating treatment from the guards. They became upset, sick, and by the termination of the experiment, many showed signs of severe emotional disturbances. The prisoners had staged riots and rebellions, and spent much of their time in schemes to outfox the guards. The guards worked continually to keep the prisoners from organizing, to prevent escapes, and quickly resorted to fire extinguishers to quell insufficiently submissive behavior. Guards forced the prisoners to count off repeatedly as a way to learn their prison numbers, and to reinforce the idea that this was their new identity. Guards soon used these prisoner counts as another method to harass the prisoners, using physical punishment such as protracted exercise for errors in the prisoner count. Several guards became increasingly cruel as the experiment continued. Interestingly, most of the guards were upset when the experiment concluded early.
>
> The prisoners weren't. By the end of the six days, many "prisoners" had asked to leave the experiment (be granted "parole") even at the cost of denial of their stipend. (This result puzzled Zimbardo; after all, the money was the major reason they stayed to that point. He concluded that the prisoners must have "internalized" their roles.) It was clear when this experiment ended that the situation, not their inherent psychological dispositions, governed the behavior of the participants; the hypothesis was disconfirmed.

The experiment was, and is, widely criticized as being unethical and bordering on unscientific. Current ethical standards of psychology would not permit such a study to be conducted today. The study would violate the American Psychological

Associate Ethics Code, the Canadian Code of Conduct for Research Involving Humans, and the Belmont Report. Critics including Erich Fromm challenged how readily the results of the experiment could be generalized. Fromm specifically writes about how the personality of an individual does in fact affect behavior when imprisoned (using historical examples from the Nazi concentration camps). This runs counter to the study's conclusion that the prison situation itself controls the individual's behavior. Fromm also argues that the amount of "sadism" in the "normal" subjects could not be determined with the methods employed to screen them. Further, because it was a field experiment, it was impossible to keep traditional scientific controls. Zimbardo was not merely a neutral observer, but influenced the direction of the experiment as its "superintendent." Conclusions and observations drawn by the experimenters, it was said, were largely subjective and anecdotal, and the experiment would be difficult for other researchers to reproduce.

Some of the experiment's critics argued that participants based their behavior on how they were expected to behave, or modeled it after stereotypes they already had about the behavior of prisoners and guards. In other words, the participants were merely engaging in role-playing. Another problem with the experiment was certain guards may have changed their behavior because of wanting to conform to the behavior that they thought Zimbardo was trying to elicit. In response, Zimbardo claimed that even if there was role-playing initially, participants internalized these roles as the experiment continued.

The larger question raised by the prison experiment, the same one raised by Milgram's study in obedience to authority and possibly more forcefully, is why on earth the participants, college students all, did not remember, and act on, their roles in the larger system, which surely should have been more solidly "internalized" over a lifetime? This was a voluntary study. Unlike German citizens threatened with loss of employment, houses, liberty and life if they disobeyed, or prisoners in the California prison system, put there against their will, these participants could have, and clearly should have, walked out at any time. They were not surrounded by barbed wire. They were in the basement of the Psychology Building. Nor were they isolated; they had each other to talk to. Why did not just one of them—especially when the study really got going and it was evident that some participants were getting sick—point out to the others that this game was not as much fun as they'd thought it would be, and that it was time to gather the bats and balls and go home to supper?

Given the universal condemnation of the Milgram and Stanford experiments, it would seem that we are not ready for the "inflicted insight" contained in both their results. We are not the people we thought we were, as individuals or as a nation. We will need a much more determined effort to internalize ethics if we are to become the people we ought to be.

Chapter 3
Allocation of Health Care Resources

How can we ensure that the riches of health care, in this the richest of nations, are allocated fairly? To begin at the beginning, what values should we be attempting to maximize in any health care system? The traditional values, put in conjunction, yield a Triple Impasse. (It's sometimes called a "trilemma," but "trilemma" is a term of logic, denoting three sentences, any two of which can be true but not all three of which can be true. We have in the Health Care Debate not mutually incompatible sentences, but objectives which cannot be simultaneously attained, leaving us frustrated. Triple Impasse seems to be a good term for it.) As used in policy debates, the Triple Impasse denotes a set of three desiderata, desirable states, any two of which are attainable but all three of which cannot simultaneously obtain. We want:

1. The highest quality of medical care available anywhere in the world;
2. Universal access to that care, regardless of ability to pay;
3. As a percent of the GDP, low and stable costs for that care.

We can have (1) and (2), but the costs will be through the roof; (2) and (3), but the health care will be bargain basement crude (*vide*: the Soviet system, through most of its history); (1) and (3), but everyone will have to pay cash, and most of the poor will not be cared for at all. Of course, the American consumer of health care internalizes all those desiderata at different points of his interaction with the health care system: (3) When he selects his insurance plan, he looks only at costs, accepting all sorts of limits to access and advanced interventions, because right now he's healthy and he wants as much of his paycheck as he can get for disposable income; (2) When he gets sick, he wants access *now*, to any part of the system, (1) When he's being prepped for surgery, he wants the best surgeon and the most advanced techniques in the country. The Triple Impasse lives in all of us, and the problem of legitimacy arises in any system that pretends to serve the will of the people but cannot attain these important objectives.

Then how shall we handle the health care costs of a population, given that the problem is already defined as insoluble? It is not too early to suggest that the attempts to place medical care in private hands (through group insurance arrangements) has not fulfilled its promise—or better, the promises that were made for it. But U.S. history predicts resistance to plans to make government the single payer, and the laudable progress in medical technology has placed high-technology medical care beyond the reach of most private budgets. In this chapter we consider a proposal to localize health care, on the model of a public school system, on the argument that such localization will answer most of the questions of legitimacy at the core of the private insurance imbroglio, provide a brake for medical costs, while preserving our ability to take advantage of the most advanced medical interventions.

3.1 The History to this Point

The oft-noted mounting frustration with the operations of the present health care system in the U.S. is the outcome of a century of development. It might be useful to see that development in three stages: Health care based in the household; health care based in the hospital; and now, health care based in the for-profit corporation.

In the first stage, until the twentieth century (in most places, until the late twentieth century), health care was simply one part of family life, like all the other kinds of care rendered to individuals by the family. As Aristotle described it, the Household, seat of the family, is the provider of the "day to day" goods needed for basic sustenance: food, warmth, basic supplies, emotional support, instruction in religion and daily living, varying in kind with the needs of the individual. Attention to health care needs is on that list. As with the other concerns of the family, family members are expected to provide the service if they can; if it is beyond their ability, they can call in some sort of medical expertise, which has been available (for a price) since at least a century before Aristotle.

The prevailing view of "health care" for the greater part of human history joined it to the human life cycle. Infants are born, have to be sheltered, fed, and, if they die, mourned and buried. Those that survive are always subject to infectious diseases (smallpox, diphtheria, cholera, malaria), that sometimes wipe out whole families. Families could do little to heal disease except bathe, feed, and comfort the patient—and watch and pray. Childbirth was dangerous, and often both mother and child died, as would everyone else, eventually. Physicians, if called, could add little to the family's efforts. Birth and death took place in the home, and were regarded as family functions. The bedside belonged to the family.

The change to the second stage began with the professionalization of the health care occupations. From the middle of the twentieth century to its end, health care emerged from its origins in the family, to become professionalized in the setting of the hospital. The rise in the prestige and authority of the medical profession coincided with the discovery of "Magic Bullet Medicine," as we came to call it. Penicillin, the first "wonder drug," was discovered in 1928, when a common mold was found to have the ability to kill colonies of bacteria. It was not widely used until World War II spurred the search for ways to cure venereal disease and infected wounds. With penicillin, we cured diseases that had killed people for centuries. Penicillin was joined by other classes of antibiotics, and by all the tools of the new Scientific Medicine: X-rays, open-heart surgeries, kidney transplants. There is no way to overestimate the psychological effect of the introduction of the first antibiotics. Every family has at least one penicillin story, just prior or just post World War II. My husband's sister died in childhood in 1938, of an infection that could have been cured with penicillin. His father contracted a lethal kidney infection in 1943, but because he was in the army, he received penicillin, and lived. I contracted a serious case of bronchitis/pneumonia at age six, just after the war, but because our hospital had been on an army base during the war, it had a good supply of penicillin, so I got well. Every family has such a list. Soon we came to believe that we could cure anything. Miracles were commonplace, and soon would be universal.

Health Care Insurance, especially as obtained through the employer, was really a product of the War: anti-inflation rules forbade wage hikes so employers offered health care insurance as a popular and widely accepted benefit. The problems of Managed Care began with the reimbursement policies adopted at the outset. Health care expenses were low anyway, and physicians, respected for their professionalism and integrity, were trusted to order only necessary tests and treatments. Insurers simply reimbursed any and all physicians for any and all treatments any insured patient received, at the physician's usual rate. Predictably, the availability of insurance drove up the frequency of visits and the cost of medical care. The quality of medical care, as measured by sophistication and effectiveness, also increased: since price presented no obstacle to complicated medical interventions, physicians treated their insured patients more aggressively, and looked for more advanced treatments to provide for them. Medical supply companies gladly subsidized physician-scientists engaged in medical research. As they found out that there was a market for any improvement they could come up with, they put more and more treatment modalities into the hands of the physician. The disparity of medical care available to the insured and that available to the uninsured began to grow. In the 1960s, the health care gap between the bulk of the American people and the unemployable elderly and very poor became politically unendurable, and Congress passed Medicare (directly administered by the DHHS) and Medicaid (administered through the states). These, like private insurers, reimbursed physicians at their "normal rate" of billing.

The high-technology Hospital was now central to all health care. Doctors no longer made house calls, not out of laziness, but because the best diagnostic and therapeutic modalities were only available in the hospital. This centralization was accelerated, if not in part created, by insurers that reimbursed only for treatments that happened in hospitals. There were two good reasons for that restriction. First, hospitals really were better monitored and regulated than physicians' offices, so there was a guarantee of quality work that could not be found for office practice. Second, when insurance first took hold, in the 1950s, patients tended to resist hospitalization, because of the pre-War association of hospitals with the dying poor.

Insurance companies reasoned that if an illness persuaded a patient to enter a hospital, it must really be serious. As an insurance claim control measure, the insistence on reimbursement only for hospital procedures rapidly grew consequences of its own: the physicians refocused their practices on the hospital, and soon could not practice without it. The high-tech hospital, in turn, came to depend symbiotically on the funds from insurance, for private payers could no longer pay the high cost of hospital care. Since the patients and the money were in the hospital, medical research and physician education soon moved in. While in thousands of communities, low-tech community hospitals continued to offer community care, the medical field was dominated throughout the 40 years after the war by large university-affiliated medical centers. The bedside now belonged to the high-tech physician and his staff in the high-tech hospital. (Popular television shows added to the drama.)

The Law of Unintended Consequences continued its work. It might have been foreseen that requiring hospitalization for insurance would eventually drive up the cost of health care, at least a little. But two results of this process were not foreseen:

1. The research that got done, possibly with some of the abuses mentioned in the last section, was spectacularly successful, surpassing the imagination of the most enthusiastic post-War boosters. Diseases that were death sentences in the 1950s (leukemia, Hodgkin's disease) could now be ameliorated and often cured. Organs could be transplanted, nearsightedness surgically cured. Surgery could be done on heads, hearts, hands, blood vessels, the tiniest infants, even fetuses, sometimes under microscopes. Gene therapy, building on scientific successes culminating in the June 2000 announcement that the human genome had been assembled, despite the spates of quackery, still holds out the promise of curing terrible diseases of genetic origin. All of this is well known, and raises the expectations of patients in the health care system.
2. The physicians have changed. With all their bills paid, and reimbursement rates continuing to rise, they soon constituted the highest paid profession. Like the businessmen and nuclear physicists of earlier generations, by the middle of the 1970s they expected, and were expected, to be aggressive, important, and very well compensated. (It is not clear that people who enter a profession with this sort of expectation can acquiesce in sudden changes of system that drastically lower their independence, prestige, and income.)

The prevailing view of health care in the half century leading to the present has changed dramatically. The notion of a "natural" life cycle has disappeared, washed away by the flood of miracles. We no longer expect to lose any infants at all. We hardly expect to die even in old age: geriatric medicine urges aggressive solutions to natural physical deterioration at greater and greater ages, even joint replacements for 90-year-olds.

What are we looking at? It seems to be a case of institutional "encephalization," parallel to evolutionary encephalization that gathered all regulatory functions of the bodies of the higher animals into the brain. As health care evolved, the Hospital, as the seat of science and the locus of scientific medicine, became central to all health care. The physician, as the scientist, became central to all decisions and the only one who could authorize any form of care, including physical rehabilitation and pharmaceuticals. There are many reasons to resist such centralization, and they seem to have occurred to everyone at once, toward the end of the twentieth century.

In the latter years of the twentieth century, we tried for a new balance, hoping to preserve the miraculous cures and the world position as premier provider of health care, while restoring the patient as author of his own fate, bringing costs under control and expectations more in line with reality, i.e. mortality. The encephalization of health care has spawned its own backlash, resisting its authority and resisting its high cost. The centrality of the physician, and the exclusion of the patient from health care decisions, ultimately offended the American preference for individual rights and autonomy, especially as expressed in the consumer movement. This movement, aiming to put knowledge and choice back in the hands of the ultimate consumer of goods and services, came to public notice with Ralph Nader's 1960 work on the automobiles of the day and rapidly spread to many types of goods and services. By the early 1970s, in all aspects of life that really interested us, the consumer was king; the medical model of paternalistic "authority mixed with condescension" was no longer acceptable. Second, the rapidly increasing cost of the system, on the mechanisms shown above, eventually became politically unendurable. Health care, in the opinion of the general polity, had become too expensive; even those willing to tolerate the present level of expenditure conceded that there seemed to be no end in sight for medical cost increases.

As the last decades of the twentieth century opened, a consensus prevailed that the successes of high-tech America were wonderful, but that they overruled patient choices and that they cost much too much: there must be a tremendous amount of waste in the system, and we must find a way to preserve the success while lowering the cost. No one doubted that this was possible. We had models. Polio, scourge of the 1920s–1950s, was one of them. Low-tech medicine had first watched children suffer and die of what was known as "infantile paralysis." Then middle-tech medicine had saved children only to spend millions on rehabilitating their crippled bodies. But then, after the discovery of the Salk vaccine, high-tech medicine had stopped the disease completely, for pennies per child. That was the model we expected to be repeated.

Starting in the 1970s, a Patient's Rights movement took legal ground (court decision by court decision) until 1990 when the right to refuse treatment and to specify treatment by advance directive was written into federal law. But even this relatively simple development has been dogged by the Law of Unintended Consequences. For with that success, as far as the bioethics profession was concerned, a "double veto" was in place for all health care decision-making: the physician could refuse, of course, to supply inappropriate medical treatment (that was entailed in his Oath, and had always been true), and now the patient also could refuse to accept treatment the physician wanted to prescribe (that was new). But that's not the way the courts read it. In a series of bewildering cases, courts have held that the right to refuse entails the right to demand. That is, since the new "right to refuse treatment" changes the locus of decision in health care from the physician to the patient, the patient must now also have the right to demand treatment that the physician does not think is appropriate. The immediate application of this change in interpretation is that if a physician deems further life support "futile" for a patient—very unlikely to lead to any recovery, and therefore a waste—and if the family disagrees, and wants the life-support continued indefinitely, the courts will side with the family, and the life support must be continued. Hospitals have tried in vain to develop policies on "futility" that would back up the physicians; such policies may be turned against them by insurers demanding an end to interventions, and the courts will not sustain them.

The 1980s continued to search for lowered costs. Early attempts to curb costs included a 1983 Medicare initiative to establish Diagnosis-Related Groups (DRGs). "Reimbursement according to diagnosis" was a Medicare attempt to determine an average stay in a hospital, and average cost for that stay, for a particular diagnosis; and a corresponding policy to reimburse a hospital, for any given patient's treatment, just the average amount determined for that diagnosis, no more and no less. If the patient's costs went over that average, the hospital had to make up the difference; if the hospital managed to discharge the patient early or otherwise to keep costs below average, the hospital kept the difference. The policy was supposed to provide the hospital with an incentive to accelerate hospital discharges. But this system did nothing to stem the increase in patients seeking treatment, an increase spurred simply by higher expectations of health among the general population. Further, as physicians discovered that some diagnoses were reimbursed much more handsomely than others, and let their patients rest in the hospital longer, diagnoses began to undergo "DRG creep"—a slow but traceable move from less expensive to more expensive groups.

DRGs were really just a tinkering with the system: like Tradable Pollution Quotas, they were an attempt to put greed to work for cost-cutting. That strategy rarely works. The logical solution to the financing problem, of course, required a single payer of all medical costs, financed by federal taxes. "Socialized medicine," as it is called, was already in place in the UK, Canada, and most Western European countries. By the early 1990s, informed opinion held that some variant of the "single-payer system" was inevitable. When William Clinton was elected president in 1992, a serious attempt was made to develop such a plan; not exactly a Single-Payer socialism, like Europe's, but at least a path to cost control that would bypass the private insurers. In the summer of 1994, after 2 years of work, it was completed, died in committee, and failed, but not before helping the citizenry to understand the issues.

Why did the Clinton plan fail? Not because of the Triple Impasse, widely recognized and accepted as necessitating compromise. It failed partly because (critics insist) there were serious errors in method in putting it together, excluding too many voices until the end of the process. It failed partly because of the publicity efforts of the private insurance companies, which really wanted a share of that tremendous flow of health-care cash, and flooded our TV sets with warnings that the Clinton plan would deprive us of valued freedoms. Ultimately the legislation failed because every special interest (physicians, pharmaceutical companies, the elderly, chiropractors, unions, women's rights movements and the Roman Catholic Church) politicked it to death in committee—each withholding support until it could get only one or two more favorable provisions for their groups. The end result, defeat of the proposal, puzzled most Americans, who honestly expected it to pass in some form. But it died, and we were left with resort to private insurance, which was, after all, the alternative recommended by the insurance companies.

3.2   Health Care in the Private Sector

Still, a widespread consensus insisted that costs had to decrease. The U.S. spends more on health care than does any other developed nation. So we embarked on an experiment to let the Private Sector manage health care. Private insurers administer health care insurance just as they do any other kind: as private arrangements for private coverage of health care needs for some, but not all, Americans. While the wealthiest of Americans continue to pay privately for their own health care, and most (but not all) of the poorest Americans continue to subsist on straight state-administered Medicaid, the majority of those in between are covered, more or less, by this private insurance system. Americans generally believe that the private sector can run any enterprise more effectively, and efficiently, than the public sector. (For those at some distance from their management training, a reminder: Effectiveness is a measure of the task getting done; efficiency is a measure of the cost, in materials and energy, that it takes to do it.)

Medical insurance is a straightforward business arrangement. Individuals or families can buy health insurance on their own, from some private insurer who offers such coverage, on an analogy with fire or homeowners' insurance. Companies can arrange to have their employees covered collectively by some insurance plan through deductions from their pay, on an analogy with a pension plan. The benefits of the health insurance plan, in either case, may vary widely: as with any insurance, the insured (or the company) can opt for a more expensive plan that covers more contingencies, or a less expensive plan that covers fewer contingencies. In any plan, the insured is issued a contract by the insurance company that is supposed to spell out what is covered, for how much, and what is not, and the insured is supposed to agree to it. It sounds very simple, and entirely in accord with our market tradition, certainly with the tradition of buying insurance: a willing buyer purchases known benefits from a willing seller, and both are better off for the exchange.

But free market benefits do not describe the citizen's experience with health insurance. Patients, the ultimate consumers, complain bitterly of varying, confusing, and changing benefits and non-benefits, periodically (unilaterally) changed by the employer or by the insurer. (Meanwhile, many Americans have no insurance at all.) The plans presume customer choice that in fact is not, in employee experience, really feasible; the benefits are difficult to keep track of. Claims seem often to result in unexpected costs in addition to the contracted reimbursement for treatments that are covered by the insurance, unexpected notifications of non-coverage for other treatments, and in delays and demands for further explanation and verification. These obstacles, bewildering to the patient, are set against our news media's ongoing accounts of the substantial profits of the insurance companies and the Healthcare Maintenance Organizations (HMO's; or similar arrangements) with their millionaire CEO's, interspersed with increasingly frequent accounts of insurance plan withdrawals from Medicare (stranding elderly patients who, with pre-existing conditions, may not be able to get private insurance elsewhere), and plan bankruptcies (which strand all their subscribers without health insurance at all), completed with stories of "severance" pay in the millions of dollars paid to ex-CEO's let go for incompetence. These stories are read against a backdrop of horror stories, anecdotes of people who thought they had good health insurance, and who were suddenly confronted with tens of thousands dollars in medical bills for which reimbursement has been denied. Underscoring the resentment is the looming threat of lawsuit or loss of credit rating if reimbursement is ultimately refused and the health care provider (physician, hospital, or home care agency) turns our account over to the collection agency. In some monstrous parody, the Managed Care experiment echoes the first years of penicillin. Everyone has a story—only this time, it is a story of terrible uncontrollable forces that take our money and then refuse us the medical care we need.

A further casualty of the Managed Care revolution may be, in the end, the healing relationship itself. The new literature on the "placebo effect" (the healing— real healing—that takes place only because of the patient's perception that healing actions have been taken) points out that the effect works only in situations where the patient is convinced that the physician is whole-heartedly committed to the attempt to treat and to heal. Not only the patient's beliefs are essential to the process. They depend on the physician's self-perception (and unconscious communication) that he is acting not out of self-interest but only to heal the patient, that he is doing "nothing that is not for the welfare of the patient," as the Hippocratic Oath put it. Let either doubt the commitment to healing, and the "placebo effect"— which can account for 20–70 % of perceived improvement, depending on the ailment—may be severely limited. And that doubt, of course, is what is sown in the minds of both patient and physician in several typical Managed Care contexts: the patient is not sure whether the physician is recommending medication rather than referring to a surgeon because he thinks the condition can be handled medically or because he doesn't want the cost of the referral deducted from his reimbursement. The physician himself is not sure whether he is prescribing the generic version of the drug because he believes it to be the best medication for the patient's condition or because the HMO will reward a pattern of prescriptions of generic drugs. From both sides, the physician-patient encounter is poisoned. It may be presumed that the trust that has always obtained among physicians—that they will make appropriate referrals and deal honestly with each other—will be similarly poisoned. Not only the physician-patient relationship, then, but also the physician–physician trust that has made the health care system work for two thousand years and more, has been eroded by arrangements that place the physician's interests and the patient's interests squarely at odds.

How did our superb health care system, the triumph of the American Century, become a source of such frustration and resentment? The answers turn not on vast conspiracies, but on the normal features of the free enterprise system that we tend to forget. Possibly a short reminder of the way the insurance business works will defuse some of the anger. Insurance is a device to spread risk so that the whole damage of an ordinary calamity will not fall on the sufferer. Since the rates collectively must cover the claims collectively, it is in everyone's interest to limit the claims. From the point of view of the companies, the demands of business competition may put them in a short-term adversarial position with the people insured. It is in their interests to write the insurance policy and collect the premiums, but it is never in their short term interest to pay the claims. A scrupulous insurance company will pay all claims that it has contracted to pay, without delay. An unscrupulous company, one focused on short term profits, will delay payment as long as possible, investing the money that it saves and reaping profits on that amount. It will also be to the company's short-term benefit, at the time a claim is made, first to argue that no settlement need be paid (hinting that fraud may be involved), then to minimize any claim acknowledged as valid, and then when an amount for the claim is determined, to delay payment as long as possible. Since payment delayed is money in the bank to invest in short-term securities, and adds greatly to the company's profits, such practices may well be justified as part of the company's fiduciary duty to the shareholders. The only reason we notice such practices more with health care than with the insurance we hold for fires and burglary, is that we need health care frequently, often as obtained from different providers or types of providers, repeatedly, over a short period of time. But we almost never have fires or burglaries.

The CEO salaries, focus of such resentment by critics of the industry, are determined competitively, as in other industries, with the choice of CEO based on projections of profits to shareholders. It is not the responsibility of the CEO, or the company's Board of Directors, if an envious public is outraged by the salaries— or, in the rare cases where the CEO does not work out and has to be separated from the company, by the severance agreements. If the CEO does not promulgate and encourage those policies that most benefit the shareholder, that CEO will be replaced with one that will. If conditions change such that the company begins to operate at a loss, the company will go out of business and its assets will be sold to satisfy creditors. If conditions change so that a large and solvent insurance company cannot make a profit on a certain line of insurance (say, health care insurance, or participation in Medicare), it will stop writing that insurance, and will dispose of the policies it has already written. These practices are necessary for business survival in a competitive environment; to that extent, they are just good business.

No insurance company could survive a requirement that it offer equal access and equal premiums to all. Every insurance company will try to insure all and only those who will never make claims, for obvious reasons; in automobile insurance, that means a preference for mature drivers, in fire insurance, that means a preference for residences with smoke alarms and commercial buildings with sprinklers, and in health insurance, that means a preference for non-smokers and people who at least are not sick already, and are less likely than others to get sick—the young, the suburban, the educated. The practice of discovering and insuring only those who will be unlikely to make claims—"skimming" (the cream), or "creaming"—is entirely acceptable. Insurance companies are even legally permitted to discriminate by age and sex—note the bias in automobile insurance against young males. So most of the horrors of managed care are simply reasonable and accepted practices in the industry, now being tried out in a new field of endeavor.

The conflicts are well known, and lethal to professional commitment. When referrals to specialists entail financial penalties for the physicians, patients are less likely to be referred to specialists. Insurers typically determine how many days in a hospital are required for each patient (i.e. how many hospital days will be reimbursed), how many days in a convalescent home, how many visits from a home care agency nurse, even when the insurance contract seems to allow for more. Where one or several insurance companies have hired an intermediary to manage costs for a certain region or area of health care, that intermediary makes its money by keeping hospital stays and physician visits below those authorized and paid for by insurance, the savings from which it shares with the hiring company. But none of these provisions are illegal. Again, it is the practice of every distributor to seek lower prices from its vendors (in this case, the physicians), and otherwise to cut its costs in the name of efficiency. There is no sink of corruption here, no foul growth of avarice and deception to be rooted out in order to obtain better health care. If we resent being treated in this way, it is probably because we are spoiled—we are used to having health care administered not as a business, but as a profession.

It may be a good idea to recall the distinction. A profession, of which medicine is the paradigm case, takes service to the client as its ethic. The Hippocratic Oath, as above, binds the physician to do "nothing that is not for the welfare of the patient." The physician must never allow his interest (as a "vendor of health care services") to displace the patient's interest in good care. Indeed, the first question I ask a physician when I consult him about my throbbing head, stomach, knee, whatever, is, do I need a doctor? Or is this something that will go away by itself, in which case I will go home and (unless the physician has had to run some expensive tests to find this out) owe the doctor nothing? If the physician put his interest first under the old fee-for-service system, I could guarantee he'd find something expensively wrong on every visit, much as I suspect my automobile mechanic of doing. I never suspected my physician of doing that, because I have always regarded my physician as committed to my good with an ethic stronger than self-interest, so reliably that it never occurred to me to think about it. This is the point at which the HMO provisions for physician compensation bear with special force: can I continue to have that trust in my physician, knowing that not only is it is not in his interest to find anything that needs treatment, but that he is under worse pressure than he could ever have been under fee-for-service: if he found nothing wrong with me under the old system, he forfeited a single fee; if he finds more wrong with me now than the HMO guidelines permit, he may forfeit his contract and the bulk of his practice. If he is a member of a group practice, that forfeiture may, ultimately, either separate him from his practice group or drag the whole group down with him. In either of those cases, he loses his entire practice. How much weight can a professional ethic bear?

The difference between the professional ethic and the market ethic, the ethic of the physician and the ethic of the business person, cannot be overemphasized: the professional is obligated to put the patient's interest first even if it means, which it regularly does, serving the patient without reimbursement; the participant in the free market, on the contrary, is not only permitted but obligated to follow that course of action that maximizes his ROI or profit, since any attempt to do otherwise in the name of charity, sentiment, whatever, can only hurt the efficiency of the system and detract from the interests of all. In the case of the corporate officer of the insurance company, the two ethics come together: the officer is under a fiduciary obligation to the corporation, to protect its interests, as the physician is under a fiduciary obligation to the patient. The corporate officer cannot, even if his charitable instincts would incline him to, authorize reimbursement that his company is not legally obligated to pay. Such authorization would be a violation of his moral duty. So the physician cannot appeal to the better nature of the insurer, or entreat him to be "ethical" in his coverage or payment policy; his ethical imperative is to protect the company. If the physician makes too many referrals or orders too many tests, it is the insurer's duty to deselect him in order to conserve the company's assets.

Nor can some version of "stakeholder theory" come to the physician's rescue, with the reassuring news that every business relationship is subject to competing claims and clashing interests, and that it is his right, and duty, to weigh the claims in each case to resolve the conflict. On a stakeholder calculation, the physician is right to balance the competing claims of the patient, the insurer, the group practice that he endangers if he drifts outside the insurer's guidelines, himself and his family, and the public good. On the Hippocratic ethic that the physician is sworn to, he has no right to do any of that. He must protect and treat his patient as the sacred duty of his profession. The professional-client relationship is totally different from the provider-consumer relationship. You can have many competing interests, but not conflicting clients. If there is financial adverse interest between clients, the professional must get out of one relationship or the other. If there is a conflict of interest between the client and the lawyer personally, the client must be referred to another professional. Any lawyer knows that.

Our trust in the physician has been based on implicit faith in the power of the professional ethic. But recall: one unforeseen result of the generous reimbursement schedules for health care practice was that we brought a different sort of physician into the profession, with a different set of compensation expectations. We are not sure that this new crop of physicians—recruited to the profession with the promise of money, deeply in debt for their education, and now finding no alternative to employment by an HMO if they wish to begin practice—will have the same resistance to financial pressures and exigencies that their predecessors had.

We are spoiled, used to being treated as trusting patients, protected by the paternalistic ethic of a health care profession. And what we find acceptable in business practice generally—the layoffs, extensive cutting of salaried positions, the streamlining to the "lean and mean" to minimize costs and maximize flexibility, the high-flying CEO's with yearly compensation in the tens or hundreds of millions—we do not find acceptable when we perceive that it is being achieved at our expense at that point when we are sick and vulnerable. To make the situation less acceptable yet, the costs of the health care system continue to rise. That is not surprising in any case—the population ages, while expectations and technology continue to rise and new pharmaceuticals continue to advance to the market. But the rise is made worse by the insertion of business. All those insurance executives and their staffs have to be paid, while all administration and other overhead costs of the business have to be met, and there is no place for the money to come from except from lowered compensation for health care professionals, from the pockets of the patients, insured or otherwise, or from the pockets of the taxpayers. Why we ever thought that spending millions to hire insurance executives to run our health-care system was going to save money may be the most puzzling mystery of the twentieth century.

The resort to the private sector, recall, was supposed to solve the problem of cost containment, one of the unsolved problems of the half-century growth of high-tech medicine. It does not seem to be working very well. But worse penalties have devolved upon the health care system. Recall that for the old and the poor, the government is still the major insurer. The government, through Medicare and Medicaid, spends one out of every two health care dollars. Seeking to recoup some of that money, June Gibbs Brown, Inspector General of the Department of Health and Human Services (that oversees Medicare) launched a new initiative in 1999, to give hospitals and nursing homes more severe supervision. Regulation of these providers to date has been hard enough. For years now, the Joint Commission for the Accreditation of Healthcare Organizations (JCAHO), a private agency formed by health care organizations, made up of accountants, physicians, and others versed in the problems of health care, has been in charge of reviewing the practices of hospitals and other health care providers (but not individual physicians). Visits have never been intentionally terroristic, in intent or in fact, but have always been terrifying, since the visit means, or is a deadline for, gathering and organizing a very large amount of information in one place at one time. JCAHO visits are always an occasion of anxiety and best behavior and, more profitably, organizing necessary initiatives—procedures that people have always known should be in place but no one had the time to develop them until the threat of inspection and evaluation was immediate. But underlying the concern there has been collegiality: essentially, the hospitals and nursing homes that are inspected support JCAHO, and no one wants anyone to "fail" the inspection.

But more recently, by initiative from the Office of the Inspector General (OIG), the collegiality is to come to an end. Inspections are to be unannounced, records are to be demanded at random, and incompetence is to be ferreted out. "Ms. Brown said researchers had estimated that 18 % of hospital patients received inappropriate care that could have caused injuries or other 'serious adverse events,'..." and blamed JCAHO for the inadequate inspections that were unlikely to detect "substandard patterns of care" or "individual practitioners with questionable skills." The Health Care Financing Authority (HCFA), that directly oversees Medicare, lost no time in calling for more frequent and unannounced inspections of all hospitals. It hinted that hospitals may well lose their accreditation if the results are not satisfactory, and that, while we're on the subject, JCAHO itself could always be replaced as accrediting agency. The result of this initiative is to add immeasurably to the hospital's troubles. After spending anxious months trying to negotiate the insurance contracts they need to keep going, and uncountable hours on the telephone trying to get insurance companies to live up to the contracts already signed, they must also expect teams of inspectors at any hour, presenting unpredictable demands for records, tying up staff and threatening de-accreditation of the hospital if they are not satisfied.

As an example and illustration of such initiatives, we may take the campaign begun in March, 1999, to recruit senior citizens in "the battle against Medicare fraud." Organized by Donna E. Shalala (then Secretary of Health and Human Services), Attorney General Janet Reno and Louis Freeh of the FBI, and launched in a series of "fraud fighter rallies" in major cities, the campaign urges Medicare recipients to call Medicare if they think they have been billed for a service that was not performed, overcharged for one that was, or billed twice for the same service. (The senior citizens are instructed to confront their provider first, but "if you're not sure, you don't feel comfortable talking with your provider or your provider's answer is not satisfactory, don't hesitate to report a questionable charge to the Medicare fraud line at (800) 447-8477.") They expected to collect a fair amount of money in this campaign; Ms Brown estimated the "improper Medicare payments totaled $12.6 billion" in the previous year. And if they do collect money, they are offering up to $1000 dollars to the senior citizen whose report led to the collection. Physicians, of course, went ballistic at the proposal, knowing well that many of their elderly patients could barely understand, let alone remember, the sets of tests and medications prescribed for them. Many patients, also, are sufficiently poor that when confusion is reinforced by the prospect of $1000, they might very well make inappropriate reports. What this newly induced conflict of interest did to the already compromised doctor-patient relationship can only be a matter of speculation. Meanwhile, the inspections of the hospitals are to be much more "rigorous." It is well within the power of the OIG to decide that unless JCAHO de-accredits a certain quota of hospitals each year, JCAHO will no longer be the accrediting agency. The prospect is not a happy one.

The sudden emergence of the Report and its figures gives rise to speculation. Why the sudden dissatisfaction? Where did their terrifying figures come from? Why the unexpected public announcement, trumpeting the result of a two-year DHHS study undertaken in evident hostility to JCAHO, after which its president is left with lame defenses of his agency that emerge only in the last part of the article? We must not forget that public agencies, especially the so-called "watchdog" agencies, have a private agenda, even as private businesses do. They too need funds, coming in the form of public allocations rather than earned income, for all the reasons that private firms do; they need to hire personnel, to secure more space, to get their work done and to deal with visitors. Above all they need to be needed. The director's salary and prestige depend directly on the amount of funding received by his or her office, which in turn depends directly on the perceived seriousness of the problem the office addresses. In the case of the watchdog agency, the temptation is irresistible to indulge in alarms and excursions to underline the importance of the office and the need for further activity (therefore funding) on its part. Is that where those numbers came from, and the accusations of laxity in which they are embedded? When agencies campaign for further funding from elected bodies, they engage, not always improperly, in "politics": the effort to influence the action of public power and the allocation of public money. Was June Brown just playing "politics" in order to advance the fortunes of her agency? Ought public officials to be held responsible for the unintended consequences of their political campaigns?

The consequences here include diminution of resources for patient care. Just as money taken out of the system to compensate the CEO of an insurance company is money taken from patients and physicians, so is time, taken from the hospital to respond to more frequent visits, more varied demands for records, accounts, and reports, is taken from the treatment of patients. Time spent monitoring is time lost, as every business knows. If the monitoring is absolutely essential—if in fact our hospitals are looting the public treasury and almost one-fifth of our physicians do not know how to practice medicine—then of course we shall have to do it. But if, as seems likely, the OIG's claims (beyond the identification of the inevitable bad apples in the industry) have more to do with our continued funding of that office rather than the health care we are receiving, JCAHO's responses to the critique will do us, as patients, much more harm than good.

A final problem in this mixed business/government system stems from its size, and from the fact that we as patients are no more likely to be unflinchingly honest than are corporate or government officers. We and our physicians (the only participants in the health care system who have not yet been accused, in this chapter, of unscrupulous action in the service of greed) are likely, when the opportunity arises, to "game the system," an expression taken originally from game theory in its military applications that comes into play whenever we deal with large bureaucratic systems with very complex rules. Rules are ambiguous and subject to interpretation. As physicians and patients discover how the rules of their insurers actually work, they learn to manipulate those rules to maximize the possibility of reimbursement for those interventions that patient and physician agree are desirable. The physician, advocating for the patient, will learn to describe symptoms in ways that make them sound more serious to the insurers' Utilization Review process. As he discovers that Utilization Review will automatically reimburse for interventions that include certain "this-is-serious" tags (vital signs must be read three times a day; an intravenous line is placed), he will order those procedures whether or not the patient needs them, in order to make sure that the patient receives reimbursement for interventions actually needed. With similar information, the patient will make sure to report symptoms that always trigger reimbursement approvals. Not everyone agrees; it has been argued that a physician's duty to serve the patient's welfare includes not only the responsibility to obtain needed medical services for the patient, but also the ancient Hippocratic duty to protect the patient from injustice—in this case, unjust refusal of claims for which the patient ought to be covered.

Even if we are not actively in the process of gaming the system, we as consumers, customers, may be expected to make the kind of rational choice that we make elsewhere in the free market. If a purchase does not seem to be in our interest, if the price is high, if we have no present need for the product, we do not buy. Just as the insurance company has no obligation to sell to any customer if the transaction is likely to lose it money, so the customer does not—or did not, until the present rewriting of the federal health care provisions—have to buy insurance if it looked like it would cost more than he will gain. And we as consumers tend not to be buying health insurance. How, after all, did we get to almost 50 million uninsured? Many of the uninsured are actually eligible for health insurance, but are unable to buy it because they do not have the $4000–$6000 (or more) that it takes each year for a healthy family of four to buy insurance, or are unwilling to allocate their meager resources to include that expenditure. Financial distress is widespread even in the best of times. With food, clothing, mortgage, utilities and car payments to worry about, and no one sick, it is very easy for a family to decide to forego the medical insurance. The choice to drop medical insurance completely can come up, after all, when you change jobs; with all the other expenses, the insurance is a hard choice. And it's like every other form of insurance: every month you stay healthy and uninsured, you congratulate yourself on having saved a bundle at no cost to yourself. Catastrophic illness, you may reassure yourself, will be taken care of, by somebody.

That reasoning is correct. Consider, for contrast, the parallel case of fire insurance. All the time that you do not buy fire insurance for your home, and do not have a fire, you save that money, and may congratulate yourself. But if you do have the fire, the loss is all yours, and it can amount to hundreds of thousands of dollars, all the price you paid for that house. So the calculation works out in favor of the insurance: since fires are few, the cost of fire insurance is low; its cost contrasts favorably with the investment you put into your house, all or part of which you will lose in a fire; and if, uninsured, you have the fire, you will bear the entire cost yourself, with help from no one. Health insurance is different. Since health needs are many, the cost of health insurance is high; you cannot set that cost against your investment in your health, because you never made any—health is free; and of course, if you get really sick, society will take care of you. In a health "crisis," the frightening kind, every emergency room in every hospital that takes federal funds has to treat you until you're medically stable: the law (The Emergency Medicine and Active Labor Act, EMTALA) says so. Nor may the hospital even ask you about your insurance until they have determined what is wrong with you and treated you until you are stable—that is, until your condition will get no worse if they stop to ask you questions. And if sickness really crushes the family finances, there's always bankruptcy, and then Medicaid will take care of all your family's health problems for the rest of your life. (If a fire bankrupts you, no one will buy you a new house.) The calculation, for afficionados of such things, is similar to the reasoning of the investors in the overstretched Savings and Loan Industry some time ago: the S&Ls promised high interest; high interest means high risk; the S&Ls were clearly not being well run so the odds of collecting were not good; but no money was really at risk because up to $100,000, by the most recent legislation, was insured by the US (FDIC/FSLIC). Investors had a lot to gain if the bonds were really paid off, and took no risk at all. The result of the calculation was an unmistakably irresponsible pattern of investment, one that cost the taxpayers billions of dollars. Health non-insurance promises to do the same.

Health care now is a (fairly) profitable business, in which vendors (health care providers) offer their services on the market to their customers (health care consumers) according to all the usual laws governing business, under the usual government regulation. As with any commodity, those without funds may not expect that the vendor will provide the service; on the other hand there is no objection to the vendor using service innovations to increase his income without limit. Whether or not he has an MD after his name, the bedside now belongs to the businessman. Since awareness of the problems of Miracle Medicine first came upon us in the 1980s, we have made several attempts to empower the patient and to contain costs. Some good has come out of the attempts—at least we can all articulate the problems by now. And not all the evil projected has really come to pass. But we have plunged an excellent system of health care into resentful confusion, and present to the world an unedifying spectacle of market greed and politically motivated "regulation" in the process of destroying ancient professions, professional practitioners and professional commitments in one blow. Can we do anything about this?

3.3 A Better Way: Return to the Local

Let's consider a new proposal. At this point, we have managed to produce a spectacularly dysfunctional situation, in which the demands that we place on the health care system are impossible to meet by any system, while the economic arrangements we have placed around ours make it very publicly incapable of meeting any demands at all, beyond the most simple. And the costs are still going up; all this resort to private enterprise, as above, has made the money matters worse. We will need a new approach—some way to make health care effective, affordable, and universally accessible. We do not have that yet.

Where do we go from here? Back to the beginning of the twentieth century when all care was given at the bedside at home and we didn't have all these high-tech tubes and microsurgeries and other extreme interventions? We sometimes talk that way, but I don't think so. Those unattractive tubes in sterile surroundings save lives and restore function to an extent undreamed of by our forebears. It isn't just that our infants live, although that development is surely pleasant. It's that after age 65, which used to be mandatory retirement because life was used up by then, our generation is taking on new careers, buying homes and, with the help of aids undreamed of by our forebears, begetting children. Our lives are really longer, more active, better in all ways. In light of the evidence, our forebears would surely find our new alleged preference for "caring, not curing" and "natural" remedies incomprehensible. We do not want to give up the benefits of high-tech medicine. Nor is it possible to go back to the autocracy of high tech of a few years ago, when our doctors could order any treatment for us and somehow, miraculously, it would be paid for. We have somehow to get health care out of that high-tech hospital—that hospital is just too expensive. We need a mid-level alternative, and there is no clear map to finding it.

Let us review. What is the central problem here? It seems to center on legitimacy. How on earth can we justify 30–40 million uninsured in a health care system based on insurance? How on earth can we explain the refusal of treatment for the uninsured when we are the wealthiest society on the face of the earth? On the other hand, how can we give away expensive health care to those who have not earned it? For background, let us remember that we recognize three major frameworks of legitimacy, none of which fit the problems of health care very well.

1. The political or public sector framework holds first priority for legitimacy, as direct provider of goods and services and ultimate regulator of all others. But we tend to be deeply suspicious of government plans to provide anything. Right now we do not accept health care in general as a government good or service (along the lines of military equipment or, until recently, the Post Office) and considering the fate of the Clinton Health Care Plan, we have backed away from making government provisions the primary rationing body of those services as provided by others (along the lines of Federal Food Stamp programs). Note that government allocation is accepted over a broad area, specifically in the entitlements that go under the names of Medicare and Medicaid. But generalization to other areas has failed, and the instability of the system often infects Medicare and Medicaid, as above: they are attacked as being too expensive, and, in a sense, not legitimate—an unjustified "Federal giveaway." They are also accused of being oversized and too bureaucratic, with some justification.

2. The free market, or private sector framework is our other major system of assumptions for legitimate transactions. If something is for sale, and I want it, and I have the money, I buy it. No money, no purchase. But we do not as a people accept health care in general as a commodity, as noticed above, to be purchased on the open market like all other goods and services. The focused rebellion against "Managed Care," sketched above, seems to prove that one beyond a doubt. But again, two qualifications need to be made on that point. First, some of the complaints about Managed Care take the form, at least, of complaints that it is not free-market enough—that full disclosure of the plan's benefits was not really made, for instance, or that there are remaining

inefficiencies in its administration. The implication of this point is that possibly Managed Care could be fixed up and perform more efficiently. Second, as with government provision, there is a large sector of health care which operates wholly within the free market, privately purchased, requiring no supervision beyond that provided by the individual health care professional, non-reimbursable and non-controversial. Some medical procedures fall in this category, all cosmetic surgery falls within this category, also the extensively advertised patent (or "over the counter") medicines we purchase for allergy and pain relief, and the services of a variety of "natural" drugs and healers, most without proven worth or medical credentials. These transactions, like the purchase of cosmetics, are only marginally in the health care system at all.

3. There is also the private voluntary or non-profit (third) sector, in which we find sizeable organizations gathered not to make money nor to carry out government mandates, but to meet some discerned need in the society by soliciting voluntary contributions from the general public and calling on volunteer labor to do much of the work. Health care for the masses started here, in the hospices and hospitals set up by religious orders or other voluntary organizations. Most hospitals remain partially in this sector, dependent upon community support as well as fees from patients, recruiting and honoring volunteer workers and donors of goods and services of all sorts. The role in health care to be played by this sector is undetermined at this time. It could be significant; it could be central. Right now, this sector is in full retreat before the profits and politics elsewhere in the system. Not-for-profit hospitals are required to take any patient who presents in an emergency and make sure that the patient is stabilized before sending him elsewhere, even when the patient is not insured and is highly unlikely to pay, on the penalty of losing all Federal funds (Medicare and Medicaid reimbursements). For a non-profit inner city hospital, this requirement means treating every victim of drugland shootings, at astronomical expense, and may lead to bankruptcy; but since the hospital is totally dependent on Federal funds, they have no room for negotiation. For-profit hospitals, that take no Federal funds, are not required to maintain an Emergency Department, so need not take such emergency cases at all, but rely entirely on a well-insured clientele to pay in full for all services. Let us rephrase that: the for-profit (private) sector is free to seek its profits where it may find them, creaming the affluent and well-insured away from the city hospitals (where part of their fees would go to support services to the uninsured), while the not-for-profit sector is squeezed by Federal law to render services beyond their ability to finance, under threat of prosecution should they refuse. Right now, in short, the third sector is helpless in the whipsaw of government enforcement and private encroachments. And no matter how empowered by new laws, it is, right now, totally inadequate to handle the incredibly expensive, incredibly urgent, needs of health care in the United States. A new role for the non-profit sector will have to be found in any future system; for the present, it cannot be the solution to the health care problem.

What have we not tried? Let us put together a utopian model that might work, if only to explore the theoretical possibilities. It is not the purpose here to argue the plausibility or political feasibility of the model. But if we find something that could work, possibly we can argue from that point to the political changes that would make it workable.

Suppose, then, in utopia, we make health care a government responsibility from top to bottom, the primary responsibility not of the Federal Government or the State governments but of the local governments, leveraged not off the high-tech hospital system but the low-tech Public Health System, on an analogy with the Public School System. As every child has a right to go to a local school, along with every other child in the neighborhood, every citizen should have a right to seek health services in a local clinic, within walking distance in the cities, open 24 hr, staffed by at least one doctor or nurse practitioner and several nurses or other helpers, entirely capable of conducting annual checkups, doing basic diagnostic tests, binding up minor injuries, and dispensing medications. The right to obtain care at that clinic would begin at 6 weeks of age; the new mother enrolls the baby at birth, and from 6 weeks on, the baby is monitored by the clinic staff while it attends the day care/primary school facility next door to the clinic. If the mother has a job to go to, she can leave the child with the day care attendants, paying for any supplies it might need; or if she has no job, she can stay and help take care of all the children of the neighborhood currently enrolled in day care. "Well baby" examinations can take place with no further disruption of anyone's schedule, since the child is already in the building; the physician has an opportunity to observe the child closely from birth and is in a good position to see what, if anything, needs medical attention. That Day Care facility continues to educate the child through primary school, till about 8 or 9 years of age; that clinic is the family doctor.

For the rest of the family, the clinic also functions as a family doctor. Open and staffed 24 h a day, it is available to take care any ailments that come up in the course of work or during the night—functioning as an "urgent care" facility. (Now, with no alternatives open, the uninsured—and the insured, after hours—land in the crowded Emergency Departments of the hospitals, often without need, when anyone with medical knowledge could have treated them quickly and sent them home.) Should a condition require more specialized care, the patient can be taken to a community hospital, but under this local system, one of the nurses from the clinic would go with them, with the patient's entire history known before some poor ED physician has to address an emergent illness in a patient he's never seen before.

It is not always evident on first hearing, why our public schools provide an excellent model for a health care system. For those unfamiliar with the system, several aspects of running the public schools in America must be kept in mind, and the parallels drawn with a utopian health care system.

1. The system is local. While Federal and (more often) State mandates sometimes set standards and basic curriculum requirements—that American History be taught at some point, that certain math and English capabilities be met by some level—it is the town that runs the schools. It will be the town that runs the clinics.

2. It is unbelievably inefficient. It is subject to every nuance of politics in the locality. It is regularly assaulted by the Gray Panthers (or other organizations representing the elderly) who want their property taxes lowered no matter how much money has to be taken out of the school system, and by the Soccer Moms and others who want the schools improved no matter what it costs. The School Boards regularly hear from organized Minorities, Conservatives, Feminists and think-tank educators all eager to dictate a new curriculum based on their special knowledge or interests for every grade, K-12. After that they have to deal with all residents of every neighborhood with a school in it, worried that planned construction will lower their property values. Constant meetings must be held to hear and discuss all these views. Constant vigilance must be exercised to keep the First Selectman (Mayor, Town Manager) from granting the school parking lot repaving contract to his brother-in-law without a fair review of bids. There is not a School Board meeting that does not leave every member of the Board convinced that there must be a better way to run things. The hostility to the public school system, based on its inefficiency and on general public perception of the same, is widespread and serious. That creates a problem for the political acceptance of the public clinic system. But the clinic system will be universal and quite adequate for the health needs of the local population.

3. It is incredibly effective, as opposed to efficient. (See above for the distinction). However inefficiently, every school district in the United States manages to put every six-year old in its district at a desk, with a book on that desk, on the first day of school. No one is not served. There is no such thing as lack of insurance. The six-year old has a right to be there, and we have not found a school district that cannot put him there. The local clinic will serve everyone, too, referring out the cases it cannot handle, but serving as the provider of first resort to everyone in the district. Problems arise later, of course. Does the high school sophomore have a "right" to a curriculum that is interesting as well as truthful? Will the clinic cover weight control? Does the juvenile delinquent have a right to a special school that will allow him to get a high school degree? Do disabled students at any level have a "right" (paid for by the town) to education modified to suit their special needs, and special expensive medications? All these questions have to be settled case by case, and are properly settled at the local level. But for pure effectiveness on the primary level, the public schools system is some kind of marvel, and we think the public clinic would be also.

4. It is legitimate. The School Board is elected in elections in which all citizens take part. The meetings are open to all. Whatever is decided, the citizens had better be prepared to live with, for they, by their participation or by their lack of participation, are responsible for it. It is theirs, and they live with the results. The Public Health Board, given charge of the clinics, would be elected in the same way.

5. It is a beautiful way to engage the energies of the citizens in the accomplishment of something real. From complaints about the inadequacy of the public school libraries, parents can be recruited to help staff them; from brilliant ideas about how the science curriculum can be enhanced, local educators can

be drafted into formulating grant proposals to purchase the necessary supplies. The public clinic will have its effectiveness enhanced by the same public participation. It is in everyone's interests to keep health care costs down, since they will be paid from taxes. When it turns out, as it does, that lifestyle choices lead to the largest number of expensive health care problems, the social pressure to stop smoking, control obesity, and watch the blood pressure may become very strong indeed—and may actually make us a healthy nation. This particular feature is more aspirational, perhaps, than analytic; but it is there. It is there because the whole enterprise is local, therefore open to participation.

6. It is easily monitored just because it is local. Corruption schemes can start easily—fake schemes to bilk the school system for undelivered gym suits or crossing guards, for instance—but since everyone is right there, able to observe what is delivered and what not, to themselves, their children, or their neighbors' children, those schemes can as easily be discovered and ended. It seems that public clinics are less subject to corruption than public schools (they should keep their eyes on the drug supply), but the same monitoring function applies.

7. There are serious inequalities in the system. Rich liberal towns give their children spectacular public educations, poor conservative towns give their children what's left over. The whole system is financed from local property taxes plus occasional state and federal grants. Anyone who cares about education knows that we must do more to even out the system's results, and clinics will suffer from some of the same problems.

8. The school system works best when supported by a national educational ethos. Read to your children. Contribute to your college. Value education: encourage your children to read and to do their homework. After encouragement in the home, with continual help from the parents, the school structure carries out the same value imperatives. The promotion of good health, and consequent lowering of health care costs, includes the same imperatives. Buy good food for the family table (preferably from plants, preferably fresh, preferably organic); teach your children to shop carefully and to cook, eat together as a family, get plenty of exercise and make sure your children do too.

9. The Public Schools System is graded by age. In every neighborhood there will be a small and simple school for the youngest, with (when things are going right) small classes and lots of individual attention. Problems that will explode in the future can be caught, dealt with, at this point. In every town there will be at least one school for teenagers, with more advanced laboratories and teachers, to engage students' curiosity and help them navigate the more complex relationships of puberty. And in every region (town, for the larger towns) there will be a high school, orienting students toward careers or preparing them for college. Take note of those careers: every high school can choose the careers for which it will prepare its students, based on employment opportunities in its own region. After high school, choices of higher education are affected by so many factors that no uniform directives would work. The post-secondary system does continue the task of regional education, in the technical school and

state-college systems. Beyond that, student interest, need, and capability direct choices to large state universities, small private colleges, or the prestigious private universities with a national constituency.

The health care system has similar branching and choices, contingent not on age but on physical condition. The local clinic is perfectly capable of treating the ordinary ailments of a generally healthy citizenry. That clinic is also the best center for organizing long-term care for the chronically ill and elderly, identifying appropriate accommodations for the handicapped, hosting the Visiting Nurses and Hospice at Home. In the ordinary case, that clinic is all a citizen would ever need, from birth to death. Should surgery be required, or treatment for an accident or serious illness, a community/regional hospital will be available; the great centers of research—Johns Hopkins, Yale, Cleveland clinic, NIH—would not function as community hospitals, but would devote their efforts to cutting edge procedures and experimental pharmaceuticals.

Then what might the health care system look like, on such a model? First of all, it would teach the value of health, and sponsor local educational programs on lifestyle, diet and exercise. It would recruit local non-profit associations (like the YMCA and the Senior Citizen Centers), government entities (like the public schools) and occasional for-profit organizations (the local Gold's Gym, for instance) to include health-oriented messages in whatever else they do. Such a campaign corresponds to encouraging parents to read to their children. Just as parental initiation of education is the best predictor of educational success, so a healthy lifestyle—watching weight, not smoking, plenty of exercise—is the best predictor of good health. Incidentally, many of the organizations that would be engaged in such an effort are already committed to it, and ask only for coordination and recognition. More than once in our lifetime, attempts have been made from the highest levels of the nation to engage the people in an ethos of good health, and results have been mixed. But they have not been total failures.

Second, it would do epidemiology, the science that the Public Health system is best at. It would keep tabs on local health—on the incidence of disease of all types, especially communicable disease—by keeping close records on the work of the local clinics, by survey, by observation; it would attempt to discover the conditions that put the people of the district at risk for bad health, and educate people how to avoid those risks. Public Health has, almost since its inception, been at once the most spectacularly successful (according to the numbers) and the most boring (according to popular interest) part of the health care system. Hygiene (there is a whole generation that falls asleep at that word), along with a safe water supply and adequate sanitation, saves many more lives than medical intervention, but there will never be an easy way to make it central to consciousness, if only because it works in the negative—health is not an attention-getter. This is why a serious educational effort must attend a move to public health based, locally oriented health care.

Third, it would create and administer an extensive in-home health care system, involving visiting nurses, social workers, health care aides, and housekeepers, with the mission of maintaining the citizens' health in their own homes and helping families care for the chronically ill and dying.

Fourth, and centrally, it would maintain those local "walk-in" clinics, a form of health care new to our era and already effective, for all primary care—well-baby examinations and shots, pre-natal monitoring, the juvenile colds and stomach-aches that plague conscientious mothers, cuts in need of stitches, falls in need of X-rays, and the normal aches and pains that plague everyone. That walk-in clinic is the resurrection of the all-purpose doctor on the corner of Main Street, diagnosing and referring serious conditions, but mostly dispensing simple prescriptions, simple treatments, and lots of comfort and advice. The assumption is that all health care problems start from here, always excepting the obvious emergencies. This local clinic is not only the locus of health care for its neighborhood, but also feeds information to the epidemiologist. The questions that an epidemiologist raises can help determine what is locally needed in the way of health care. Why, for instance, is the clinic seeing so many cases of juvenile asthma? Why are so many line workers from the local chemicals plant coming in with raging headaches? Is there something wrong with the emissions from the recycling plant, or the air in the chemicals plant? At least the start of community research might be here.

Fifth, it would maintain one or more elder daycare centers and nursing homes, with varying levels of skilled nursing care according to need, for the elderly and chronically ill who cannot be maintained, at least not all the time, at home. The nursing home might also offer respite care for caregivers who need time away from the homebound patient. The daycare centers would be extensions of the Senior Centers presently funded by most towns, providing health-related services and recreation for independent seniors. (Creative towns put them in the same buildings as town childcare centers, and let the separated generations amuse each other.)

This local health care system would be governed, like the local elementary schools, by a local board, the Health Care Board, and would be funded by local taxes. There might very well be parallel systems to the public health care system. Religious denominations might wish to set up their own clinics following their own teachings on health care, as they set up their own private schools. Private associations might wish to set up private health clinics restricted to an elite clientele willing to pay high premiums for the privilege of specialized or cosmetic care and no queues, along the lines of elite private schools. But use, by choice, of these private or parochial clinics would not exempt patrons from full payment of the taxes needed to support the public system.

Should an acute problem go beyond the X-rays and stitches capabilities of the local walk-in clinic, each large town or region would include a secondary care facility, or community hospital, capable of most surgery, obstetrics, and treatment of most acute diseases that require hospitalization—pneumonia, for instance. This facility could also care for the very seriously chronically ill, and for the dying who cannot be cared for at home or at the local nursing facilities. Whatever the origins of the hospital (religious order, private voluntary, municipal), the hospital would be governed by a Board drawn from the local Health Care Boards, and funded by a combination of state taxes, levies from the towns, and private contributions.

For the high-tech medicine for which the United States is famous—the microsurgery, gene therapy, organ transplants and miracles yet in the process of discovery, a nationwide tertiary (and quaternary—a term first heard this last year) care system would be maintained. Most of it is in place now: multipurpose centers for medical education and research as well as treatment of the most unusual cases, sponsored jointly by private foundations and the federal government, more on the model of the National Institute for Health (NIH) than on the model of the community hospital. These centers would not provide health care that could be provided at either of the lower levels. They would be governed by a combined board of professional hospital administrators and federal officials, funded by Congressional appropriation.

The agenda of this multi-tiered system is transparent: It aims to provide as much skilled care as is needed to prevent health problems from happening and keep them close to home when they do. It aims to miniaturize high-tech medicine (ventilation, dialysis, chemotherapy) to be used appropriately in local walk-in clinics and in the home if necessary. It aims to bump the authority and capacity to manage health care back to the local level and the local political process, which by its very inefficiency, makes sure that all voices are heard. It aims to create a health care system which is just, legitimate and effective—easy to monitor, from which no one is turned away, in which all participate as a matter of right, which preserves the ability to carry on high-tech medicine, and which limits costs by folding health care in with the other functions of the local community, where it belongs.

The sophisticated medical care of the high-tech hospital is retained in this scheme, but it is not the centerpiece of health care as it is now. The centerpiece is the education in preventive medicine, sanitation, environmental medicine and epidemiology—the domain of the Public Health Service. As minor ailments crop up, they can be treated locally and the "patient" (for as short a time as possible) returned to normal activity. As age and disease make normal function impossible, long-term care can be offered in the home, or, if necessary, in the local nursing facility. If the facility becomes full, the town will do what it does for the public schools—find portable facilities until the facility can be expanded, and that right early. The option of refusing care to a local person is simply not in the plan. It is possible—indeed, it should become the norm—that the local health-care facilities are all that most of the citizens need for most of their lives. It is the norm, after all, that students in the public schools do not need remedial reading or specialized individually tailored education plans.

The goal is simple. We should be able to say, and see, how a single child will be schooled from pre-school through college; we should be able to say, and see, how a single individual would be followed by a single health care system for a lifetime. At every health juncture, recourse is had to the least restrictive, expensive, and invasive treatment, closest to home.

With the adoption of some such plan as above, the community reclaims the bedside. The family can no longer reclaim it, simply because health care has gone too fast and too far and we are not willing to give up any significant amount of its benefits. The physician (as the physician is well aware) is a poor guardian of the bedside; the medical arts and science do not contain within themselves the science of their own limits. But the businessman is much worse, and not aware of it; we could wish that we could hear the same professions of humility from the business community that we heard from the medical community in the 1970s. Health care simply has to be got back in proportion and in the hands of health care professionals to run for the benefit of the citizenry, not the stockholders. A community-based health care plan can do this.

Now, why is everyone so sure that it could never happen? Let us count the ways, attempt answers briefly, and leave the rest for discussion and discovery in the experiments to follow.

First, there is the process inefficiency. The running of the public school system in any community takes far more effort than seems to be expended on any other system. Our love for the free market system comes from its ability to satisfy our needs invisibly, without public hearings. The public sector can do that only under the best of dictatorships. With the public schools, or the public health, we will have nothing but meetings and arguments that seem to go on forever. In the end, the battle goes not to the strong, or the ones with the right answer, but to those who have the most time to go to meetings. But: the legitimacy comes from this aspect alone. It is inefficient because all voices are heard, many times, which means that the end is legitimate and just. We could run our public schools more efficiently, if we really cared to try; the interest in keeping the system open outweighs the interest in making it work efficiently. (The interests in an efficient health care system might just tip the balance the other way.)

Second, there is the resource inequity, from community to community, an inequity sure to be duplicated in any health system. To be sure: a rich community will take care of its people well, a poor community will be unable to do the same. This is not fair. There are no easy answers to this objection, any more than there are easy answers to any claims of injustice stemming from the inequality in capitalist countries. All that can be urged is that the health care system, like the public school system, is not entirely dependent on money to work well; concerted effort on the part of the citizens will improve both, multiplying the effect of what money they have. And quality in the health care system, like high SATs in the public school system, will raise property values as the community becomes more attractive; that in turn will raise the tax revenue and the general wealth of the community. And meanwhile, within the community, there will be no one who is not served.

Third, there are the political inefficiencies, in this case ignorances, introduced in the political process. Both education and health care require expertise. There is no way to guarantee, in the political process, that expertise will be put in charge. By way of answer, it may be conceded that there is no guarantee, but common sense and ordinary wisdom suffice to put well qualified persons in charge of both systems. State and Federal requirements, always necessary, can constrain those communities temporarily at the mercy of organized ignorance to recruit qualified professionals to run both educational and health care systems.

Fourth, there is the parochialism in subject matter choices, in the educational system, readily translated into the physician-preferred practices in the medical system, in which local preferences and prejudices have sway. As with the educational system, state and national oversight can reduce the worst of the problems foreseen. It is not likely that the health care for an entire town would be turned over to the Christian Scientists or the aromatherapists, even if they should have a political majority. But beyond the worst abuses, the very local emphasis of the health care system is one of the advantages. Different locales have different health care problems, and should be allowed to individualize emphases in health care provision. The United States is a very large country, and it is not clear that a one-size-fits-all National Health Care System would be able to take account of differences in physical and cultural environment that affect health.

Fifth, there is the serious obstacle to the adoption of any such system, that it will get in the way of certain interests continuing to make money. Insurance companies would chafe under local rules; pharmaceutical companies would not have a national market; physicians and other professionals may very well resent being hirelings of a town; and managed care intermediaries—the firms that achieve savings for insurance companies by keeping utilization below contract—would have no further employment. Any entity that has taken up residence in the enormous cash flow of the health care system will oppose its eviction. When the Clintons' health care plan was in debate, we saw the enormous effectiveness of just one sector of the economy—the private insurance sector—in turning public opinion against the plan. How much more vigorous would be the campaigns against the turn to the local! There is no easy answer to this objection either. The forced decentralization of health care decision making power will interrupt the profits of national companies and the power that they exercise over practitioners. For answer to this very practical objection, we rely on the perception that everything else has been tried, and has not worked, that the system really is less expensive over all without the huge insurance establishment to support, and that local control at least gives us local control.

Sixth, and most seriously, the people of the U.S. have repeatedly demonstrated a willingness to purchase for themselves as participants in the private sector what they have been unwilling to purchase for themselves collectively as citizens through their taxes. John Kenneth Galbraith first noted this in his groundbreaking *The Affluent Society*, a general commentary on Post-War America. We would buy the most expensive cars to run on the worst roads; we would rather carry around a private oxygen tank than pay to enforce clean air regulations. We would rather buy private school admissions for our children than work to improve the public schools; it is likely that we will always prefer to buy health care for ourselves than work to improve a local health care system. Without support, how can such a system arise and survive? Again, for this objection, there is no easy answer. The only answer at all may be the one offered above, that the hunger for a health care system that works, and can be monitored locally, may overcome our unwillingness to entrust anything to the public sector.

Locked into our private sector free market syndrome, we have seen the future of health care, and it doesn't work at all. We cannot go back, not to the bloodletting, herbs and prayers of 150 years ago, nor yet to the spare-no-expense exuberance of 50 years ago. We cannot stay in the bewildering morass of conflicts of interest created by the present experiment in "managed care." A National Health Care System seems like a logical solution, but it does not seem to work here. What's left? Let's give the local community a try.